ENERGY SOURCES AND POLICY

AN OVERVIEW AND GUIDE TO THE LITERATURE

By ISRAEL BERKOVITCH

Energy Sources and Policy: an overview and guide to the literature
ISBN 0-7123-0805-9

Published by:
The British Library, Science Reference and Information Service (SRIS)
Marketing and Public Relations Section
25 Southampton Buildings
London WC2A 1AW

British Library Cataloguing-in-Publication Data
A catalogue record for this book is available from The British Library

Desktop publishing by Tony Antoniou.

For further information on SRIS titles contact Tony Antoniou on 0171-412 7471.

Front cover illustration courtesy of British Nuclear Fuels plc (BNFL), 'Thermal Oxide reprocessing plant (THORP)', Cumbria, England.

Acknowledgements

The author would like to thank all the organisations mentioned in the *Guide* for the use of their material, in particular: HMSO, BP (The British Petroleum Company plc), Shell International Petroleum Company Limited, APPE (Association of Petrochemicals Producers in Europe), OECD (Organisation for Economic Co-operation and Development), WEC (World Energy Council), Kogan Europe, Energy Savings Trust, British Gas, IT/Power and the NRPB (National Rivers Protection Board).

Contents

Foreword i

1. Energy Trends 1

2. Coal 19

3. Petroleum 37

4. Gas 55

5. Nuclear 69

6. Renewables 91

7. Conservation 113

8. Policies 131

Organisations Connected with Energy 151

Glossary 165

Index 169

intentionally left blank

Foreword

Energy is a topic which is rarely out of the news. It is also a broad and complex subject which crosses the disciplines of the physical sciences, environmental studies, economics and politics. Knowledge and understanding are vital for informed debate, yet the layman will find the technical literature daunting. Even the specialist in one area may find it difficult to master all the related areas.

With its wide-ranging introduction to the subject, *Energy Sources and Policy* can help those with a professional interest in energy and those simply wanting to understand what is behind the headlines. Separate chapters cover each of the main energy sources, including: coal, petroleum, gas, nuclear and renewables. The book provides brief histories of the different energy industries and examines the technical, economic and 'green' issues associated with each.

It also considers approaches to the problems from UK, European and international perspectives. After a chapter on energy conservation, the concluding section of the book draws together all the strands in a discussion of world energy use, set against the background of the 1992 Rio de Janeiro conference on environmental issues.

Much of the literature cited in the references and bibliography is available for consultation in the Science Reference and Information Service Reading Rooms or to registered users of the Library's Document Supply Centre. For more information about British Library services please telephone 0171-412 7473.

Dr Israel Berkovitch is a chemist whose career spans employment as a scientist with the former National Coal Board and as a Technical Director at Polycell as well as working as a writer and lecturer. He is author of *Coal, Energy and Chemical Storehouse* (Redhill. Portcullis Press, 1978) and of the 'Environment' section in recent editions of the annual *Pears Cyclopaedia,* besides being a regular contributor to technical and general magazines on energy and popular science topics.

In a rapidly changing field new issues are constantly arising and new publications appearing. The book is up to date as of Spring 1995, when the author completed work on the text.

The views expressed in the book are those of the author and not necessarily those of the British Library.

The British Library Science Reference and Information Service
October 1995

intentionally left blank

Chapter 1

Energy Trends

Contents

- Environment and Energy
- Some World Trends
  - Increasing Population 'drives' Energy Demand
  - Energy Intensity
- Patterns of Energy Supply and Demand
  - Price Subsidies Criticised
- Reserves
- Trends in the European Union
- UK Trends
- What is the Pattern of this Review?
- Footnotes
- References

Energy Trends

"Energy is the only life....Energy is Eternal Delight"

from 'The Marriage of Heaven and Hell' by William Blake

Though Blake was referring to energy as being 'alone from the body', humans have ever more transcended the limited abilities of these sources. By drawing on sources of inanimate energy, people have both reduced the need for heavy manual labour and improved their material living standards. Worldwide each of us now consumes on average a little over the energy of one and a half tonnes of oil per year.

We also know that this average figure conceals a wide variation between different countries and different social groups within countries. Using the same units (average tonnes of oil equivalent per head per year) energy consumption ranges from about 7¾ tonnes in North America through 3¼ tonnes for OECD Europe to about ½ tonne for the rest of the world. [1]

World Consumption

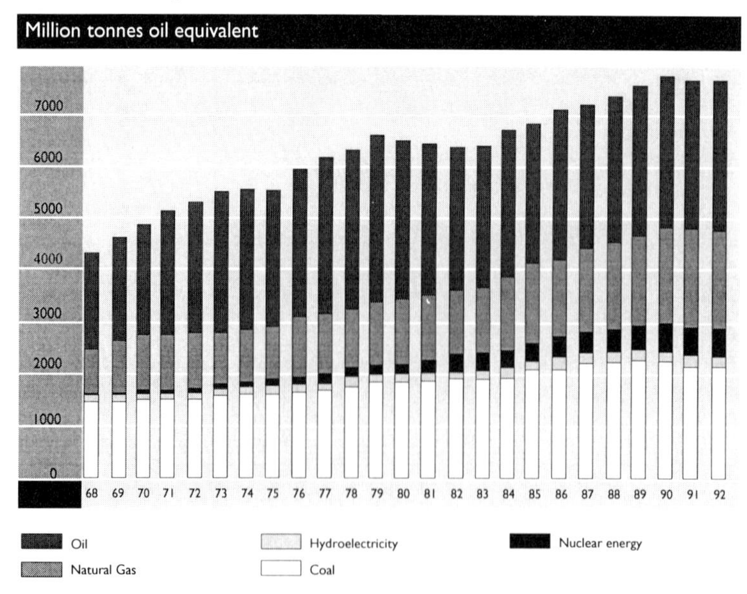

Source: BP Statistical Review of World Energy 1993

Broadly, those using the highest levels of energy have the highest standards of living. Higher levels of energy consumption have been shown to be linked to health and social benefits including higher life expectancy, lower rates of infant mortality and lower levels of illiteracy. Such facts lead to the truism that consumers value energy not for itself but for the services that it can provide, such as heating, refrigeration, cooking, lighting and motive power. Ironically, it is the people who most lack the benefits of higher energy consumption and the associated industrialisation, who bear many of the environmental costs of these developments elsewhere.

The point of re-stating these relatively well-known features of world energy use is that developing countries are well aware of them and are fighting to raise their living standards by measures that will result in using larger amounts of inanimate energy. This is already a marked trend. Though world (commercial) energy use rose by only 0.2% in 1992 compared with 1991, this average too hid significant differences. Countries grouped as 'Non-OECD Europe' – mainly the former Soviet Union and countries in Central and Eastern Europe – showed a fall of 7.7%, while the countries grouped as 'the industrial countries' had modest growth of 1%. However, in the less developed countries (LDC) demand rose by 4.8%. Within this group were some startling rises such as those for The Philippines (12.8%), South Korea (12.2%) and Thailand (10.2%). [1]

Environment and Energy

In the last two decades there have also been growing pressures for attention to the environmental effects of all human behaviour – including, naturally, the use of energy sources. These pressures culminated in the far-reaching UN conference in Rio de Janeiro, Brazil, in 1992, which has had a major influence on national energy policies throughout the world. Its formal title was the United Nations Conference on Environment and Development, mercifully shortened for reference to 'The Earth Summit', and it convened a record number of 103 Heads of State and Government for its two-day Summit Segment.

Preliminary documents indicated a recognition of the interaction of many factors in ensuring 'our common future' and the need for a global partnership between developing and more industrialised countries. So although the first area listed was protecting the atmosphere (climate change, the ozone layer, air pollution), it went on to refer to such matters as deforestation, protecting oceans and using resources rationally. The Earth Summit adopted a group of documents with the aim, in the words of the UN Secretary-General, Boutros Boutros-Ghali, of creating 'a new mode of civic conduct'. They were:

- The Rio Declaration on Environment and Development, a set of 27 principles to govern behaviour of individuals and nations in the quest for global sustainability;
- Agenda 21, a blueprint for action for sustainable development of the planet;
- A Statement of Principles on managing, conservation and sustainable development of forests.

In addition, two Conventions that had been in preparation for some time before the conference were 'opened for signature.' By the end of the proceedings both had been signed by 153 States. These further documents were:

- The Convention on Biological Diversity to ensure effective national action to curb the destruction of biological species, habitats and ecosystems;
- The United Nations Framework Convention on Climate Change[2]. The Convention enters into force after ratification by 50 'States parties' (that is parties to the Convention agreement). The aim is to protect the atmosphere from a build-up of gases generated by mankind ('anthropogenic gases') that trap heat from the sun causing an enhanced 'greenhouse effect'. This is the effect akin to that of the glass in a normal greenhouse, allowing heat radiated from the sun at shorter wavelengths to pass inwards to the Earth, but absorbing part of the infra-red radiation of much longer wavelength radiated outward from the Earth. In its submission[3] to the House of Lords Select Committee, the Royal Society (the UK Academy of Sciences) pointed out that the present atmospheric burden of greenhouse gases traps about 40% of the radiation from the Earth's surface. Because of the minor constituents of the atmosphere – water vapour, carbon dioxide, methane, nitrous oxide and ozone – the mean temperature of the Earth's surface is 32 K warmer than it would be otherwise.

Since the beginning of the industrial revolution, owing to human activities, the concentration of carbon dioxide in the atmosphere has risen by about a quarter and that of methane has about doubled.[4] Reports from the Inter-Governmental Panel on Climate Change (IPCC) indicate a general scientific consensus internationally – though with some well-informed dissidents – that these major changes are likely to lead to global warming with potentially damaging effects.[5] The IPCC Secretariat was set up in 1988 by the World Meteorological Organisation and the United Nations Environment Programme 'to carry out a regular assessment of the climate change issue providing the basis for development of realistic and internationally accepted strategies for addressing climate change.' Though the other gases also play an important part, the most important greenhouse gas is carbon dioxide and this has been increasing in concentration largely due to humans burning fuels. In the UK in 1991, for example, it was estimated that over 95% of carbon dioxide emissions came from fossil fuel combustion. So this issue is of key importance in developing energy policy.

The Climate Convention set out, as an objective, stabilizing concentrations of greenhouse gases in the atmosphere at a level that would prevent dangerous interference with the climate system and this level should be reached 'within a time-frame sufficient to allow ecosystems to adapt naturally to climate change.' Practical steps were specified to achieve this objective.

The developed countries, which were named in the text were due to return to their 1990 levels of greenhouse gas emissions by the end of this decade. Some of them were also committed to provide new and additional resources to help developing countries to meet their Convention commitments and to help their development along sustainable lines. This would include transfer of new and energy efficient technologies. Under this Convention there is due to be a Conference of Parties at least twice before the year 2000 – the probable dates are thought to be 1995 and 1998 – to review the implementing of these commitments, as well as amendments. This Conference will also take decisions about further steps and emissions after the year 2000.

At the end of 1993 the UK ratified the Convention. In the government report ('Sustainable Development' [6]) on the UK strategy for implementing the Rio decisions, 'the programme starts from the premise that sustainable energy use is the responsibility not just of government, but of everyone. The Government's role is to provide the correct fiscal, regulatory and financial framework for the programme.' After producing a range of possible scenarios[7] the Government has selected a representative one. This shows the need for the UK to reduce projected emissions of CO_2 by about 10 million tonnes of carbon (10 MtC) in 2000 to meet the Convention commitment. (This is about 6% of the estimated anthropogenic CO_2 emissions by the UK, which in turn is about 2.6% of world CO_2 emissions from burning fossil fuels.) The overall figure for proposed UK reduction comprises 'expected reductions in emissions' of 4 MtC in respect of energy consumption in the home, 2.5 MtC in that by business, 1 MtC from the public sector and 2.5 MtC reduction from transport.

Some World Trends

After a major worldwide study involving the recent history of energy usage, but also social, demographic and other relevant factors, the World Energy Council (WEC) has published an analysis of energy for tomorrow's world [8] that is examined in the final chapter. First, however, we must look at recent and current trends – many of them reported within the WEC study.

World consumption of energy is documented by the *BP Statistical Review* [1] in terms of individual fuels as well as the overall totals for primary energy noted above. It reported that world consumption of oil grew by just 0.5% in 1992. The pattern was much the same as that for the overall energy total – demand grew strongly in the less developed countries (LDC), grew modestly in the OECD, and fell in the countries of the former Soviet Union and those of Central and Eastern Europe. Fastest growth was in Asia where the peak rate was shown by South Korea.

A similar pattern was indicated for consumption of gas, even including the placing of South Korea in the star place for rate of increase. In the LDC gas continued to gain in usage displacing other fuels. But world coal demand was almost unchanged compared with that of the previous year, though the longer term trend is of a slow increase in use. Both features contrast with the continuing fall in demand for coal within the UK which is discussed further below. Generation of both nuclear and hydro-electricity fell a little in 1992 compared with 1991. In both cases, this decline followed a long run of annual increases shown in this source as being at least from 1966.

What is of enormous importance in assessing both these past trends and the prospects before us, is that the World Energy Council (WEC) estimates that at present some 50% of the world population do not have access to commercial energy.[8] The overwhelming majority of them – more than 2.5 billion people – have no energy other than muscle power, their own and that of domestic animals, plus traditional fuels or energy sources. These may include fuelwood, animal dung, crop residues, charcoal, or peat, and using in simple ways sun, wind or water power. In many areas, notably in Africa, finding fuelwood has become so much more difficult that it is referred to as the 'fuelwood crisis'.

Fossil Fuels R/P Ratios at end 1992

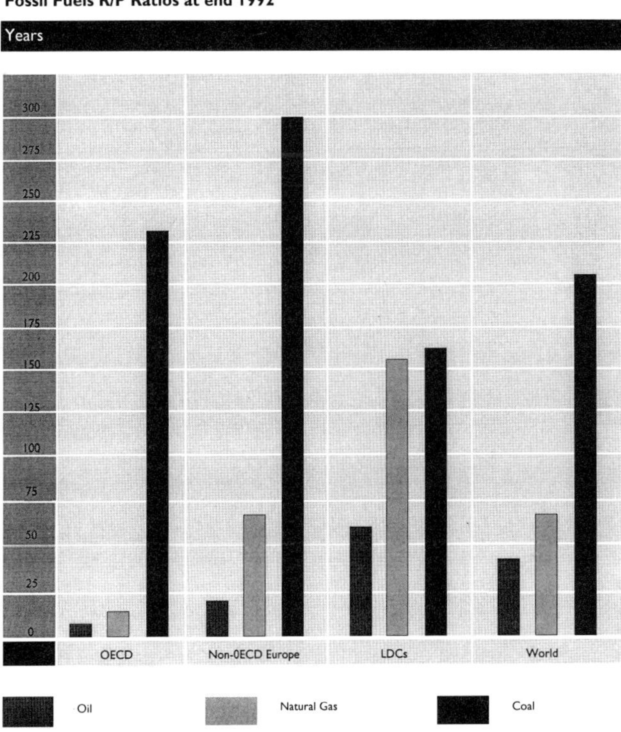

Source: BP Statistical Review of World Energy 1993

Increasing Population 'drives' Energy Demand

'The two main forces driving energy demand in developing countries have been population growth and economic development' states the WEC. Further important factors are the trends in life expectancy (which is increasing) and in the efficiency of using energy.

Growth in energy demand is already fastest in the developing countries, where it rose by 49% in the last ten years compared with 14% in the developed countries. It is also widely known that the developing countries have accounted for a very high proportion of world population growth in the past thirty years. Rates of growth of population have been generally slowing in most countries. Yet these rates of growth are still at levels that led to the organisation of a unique 'Science Summit' on population. This was a conference of representatives of world academies of science[9] in 1993 in preparation for the UN International Conference on Population and Development in Egypt (Sept 1994). In a carefully balanced, yet urgent, statement the Science Summit organisers reported that the evidence did show the developing world with its problem zones; yet there had also been in those countries some spectacular population programmes 'which were models of how to tackle all the interconnected issues'.

Energy Intensity

What are the trends in efficiency of using energy? A valuable criterion is the 'energy intensity'. This is the proportion of energy used to the Gross Domestic Product at constant prices. So this is an overall measure of changes in energy efficiency in a country or an area. Analysis of records indicates that there have been big increases in efficiency in producing, converting, transporting and consumption of energy since the start of the industrial revolution. For the UK, USA and European countries the improvement started earlier and for Japan it is reported as developing after 1950.

If energy intensity is plotted on a graph against time it becomes evident that not all countries follow the same path. Scientists and technologists do communicate with each other, so countries that industrialise later seem to benefit from the knowledge, skills and technologies of the earlier ones. This should be very relevant to the probable process in developing countries, which have the further benefit that they are generally warmer and will need to use less energy for space heating. Other factors are also involved. For instance introducing energy-intensive heavy industry (as part of what are called 'structural shifts in the economy') will apparently make the energy intensity worse, but productivity gains that raise the value added per unit of energy improve the energy intensity. The WEC notes that progress of energy efficiency is intertwined with many factors affecting demand; it quotes a definition for improvement of energy efficiency 'as any action by a producer or consumer of energy that reduces the use of energy without affecting the level of service provided'.

Patterns of Energy Supply and Demand

Commercial fossil fuels are estimated to have supplied over three quarters of total energy worldwide in 1990. Non-commercial sources such as fuelwood and dung — the most difficult component to assess — contributed 11%, nuclear 5%, hydropower 6% and what are called 'new' renewables 2%. The latter, in contrast to the traditional 'biomass' (material of biological origin such as fuelwood, crop residues and dung) are the more modern applications of solar, wind, geothermal, ocean, small hydropower and the modern controlled supply of biomass.

But what have been the trends in demand?

Here, the researchers found that transport and electricity are the two fastest growing sectors. In fact, the increase in road transport is noted as a major contributor to the rise in oil demand. All forms of transport take about 30% of the energy used by final consumers. In the industrial countries road transport was responsible for 80% of this sector, with air transport taking 13%. Road transport, overwhelmingly using oil products, is estimated to yield 14% of the CO_2 from the burning of fossil fuel. Continuing increases are expected in road transport in all countries but, in the developed countries, there is perhaps a favourable omen in the fact that the rate of increase in energy demand for road transport has declined. The WEC researchers even expressed some hope that 'saturation levels may operate at lower levels than sometimes projected'. The analysis of trends is accompanied by comments on environmental damage and the suggestion that road pricing might be applied and seen as a use charge rather than a tax with revenues invested into collective modes of transport and schemes for enhancing the environment.

The other fast-growing sector of energy use, electricity, has almost doubled in its proportion of total energy over the last thirty years. This share is increasing in all parts of the world but again at a declining rate in the OECD countries and those of Central and Eastern Europe, yet growing fast in developing countries. Fuelling this growth were all sources — escalating quantities of coal, oil, natural gas, nuclear energy, large hydro units and the more modern versions of renewable energy.

Price Subsidies Criticised

Both the World Bank and the WEC have severely criticised the fact that commercial energy prices are subsidised in many countries, notably in what are called the 'economies in transition' (the former communist countries) and the developing countries. Presumably this policy of supplying energy at prices below the costs of production and distribution has been based on the view that this would help poorer consumers (usually summarised in the phrase 'for social reasons'), sometimes venally as a political favour or more virtuously to promote faster industrialisation. But the World Bank report[10] states that these subsidies waste capital and energy resources on a very large scale. As a result, developing countries are said to use about 20% more electricity than they would if consumers paid the true marginal cost of supply. In turn this discourages investment in new, cleaner technologies and processes that are more energy efficient. This implies that assistance given to poorer fuel consumers should involve other social measures than subvention of fuel prices.

In agreement with this World Bank analysis, the WEC infer that 'prices which cover production costs and externalities [that is environmental impacts] are likely to encourage efficiency, mitigate harmful environmental effects, and create an awareness favourable to conservation.' The harmful environmental impacts include those of dust or grit and sulphur dioxide as well as the CO_2 emissions. Generally, the developed countries have priced energy at their economic prices, though these have excluded the externalities; atmospheric pollution has been greatly reduced in recent decades and further progress is expected. Of course, public opinion in the developed countries has pushed governments into introducing clean air laws and smokeless zones to force this tendency ever faster. Improved combustion methods, changes of energy source, coal cleaning, and improved styles of combustion can reduce emissions further. For coal these improved styles can include 'fluidised' beds where the coal is burnt on a bed of fine ash with air blown through it. Combustion is at a lower temperature and there is lower production of oxides of nitrogen (known as NO_X). Injecting such materials as limestone into the fluidised bed can 'fix' oxides of sulphur preventing them from being discharged into the atmosphere. On this issue, the WEC conclusion is that 'the diffusion of more efficient and more modern technology is probably the most cost-effective contributor to curbing CO_2 and other greenhouse gas emissions from fossil fuel combustion, and the most favourable precautionary environmental measure at hand.'

Reserves

Fossil fuels exist in finite amounts in the earth. Prospecting goes on more or less continuously and the proven reserves are often re-assessed in relation to rates of production. There are a number of ways of defining what is meant by the term reserves but if the concept is restricted to 'proved reserves' this is taken to be the quantities estimated to be extractable under current operating and economic conditions. If proved reserves are considered in relation to current levels of annual production, the reserves of coal (at 232 years) are due to last more than twice as long as the combined reserves of oil (43 years) and gas (65 years).

What sometimes seems strange is that despite generally increasing consumption, for many decades these ratios of reserves as a multiple of annual production rates, have increased due to the continued searching for these precious fuels. For oil, the proved reserves more than doubled between 1967 and 1992. Most of the major increases have been in the Middle East, although there have also been major finds in the last few years in Venezuela. Over 65% of world proved reserves are in the Middle East, which explains a large part of current foreign policies of the major oil consuming nations. The figures quoted in the BP Statistical Review exclude the reserves of shale oil and tar sands which are thought to have great potential, though with many problems, for the future. These shale oils are the oils obtained by heating certain shales, containing a material called kerogen that decomposes on heating distilling off the oil. Tar sands are similarly deposits that require processing in order to yield the oil. For both these sources, there are processing costs and concerns about environmental effects.

Proven reserves of gas increased about fourfold between 1967 and 1992 the main growth of reserves was in the former Soviet Union and the Middle East. These two areas in any case already have 70% of the world's reserves of natural gas. Coal reserves are more evenly spread throughout the world but some of the larger nations are reasonably well-endowed. North America contains about a quarter of proved coal reserves, the former Soviet Union almost a quarter, and China 11%.

However, despite the continuing increases in proved reserves, referred to earlier, the WEC investigations led to the view that 'the finite nature of oil and natural gas reserves will become abundantly clear within the next century'. One of their 'energy Cases' of assumptions for the future that is discussed later, leads to estimates of global energy demand that will cause pressures on the availability of oil and natural gas within a few decades.

Trends in the European Union

A marked feature of the energy pattern in the European Union (EU, the re-named European Community EC) has been the decline in the coal industry. Between 1991 and 1992 the fall amounted to 8.4 million tonnes (4.3%), and in the following year there was a further fall of 26 million tonnes (14%) to 158.6 million tonnes in

Energy Consumption per Capita

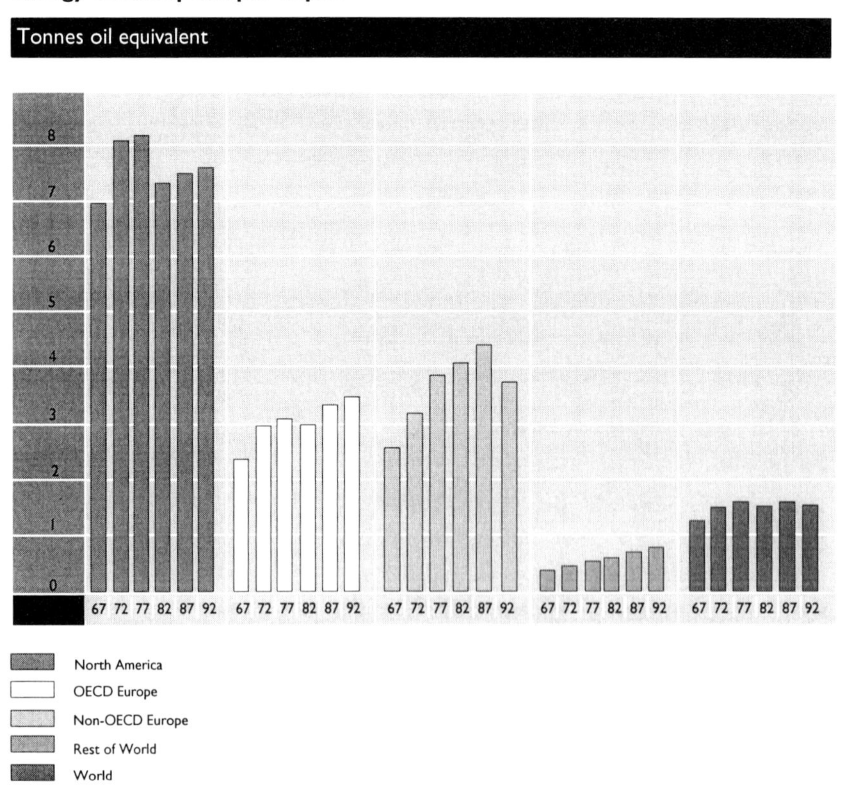

Source: BP Statistical Review of World Energy 1993

1993. The EU depends on imports for almost exactly half of its energy supply and its overall consumption of energy has slowly increased in recent years. But the composition has been changing — coal declining, oil use stable and gas use rising. For 1992 the following figures were given[12] for total consumption of each energy source:

Source	Per cent of total consumption
Hard Coal	15.6
Lignite	5.2
Oil	43.5
Natural Gas	18.6
Thermal energy produced in nuclear power stations	13.6
Renewable energy	3.5

A welcome sign of the times was the announcement by the EU early in 1994 for the first time of figures on the contribution by renewable energy. Production from renewable energy sources in 1992 — solar, geothermal, wind, hydro, biomass (wood, wood waste, municipal solid waste and biogas) — was equivalent to 43.5 million tonnes of oil (Mtoe), representing 6.7% of total primary energy production. Hydropower formed 8.2% of total EU electricity production, with France as the biggest producer. For households, biomass — particularly wood — was estimated to have contributed 7% of their final energy consumption and Denmark received a mention as the biggest producer of electricity from wind energy, although it represented only 0.06% of EU electricity production.

In the European Energy Policy proposed by the EU Commission[13], there is an analysis of prospects for energy supply and demand trends to 2005 with a brief look ahead to the middle of the next century. The policy framework is based on three fundamental aims: —

- Developing an internal market in energy;

- Developing external energy relations and ensuring security of supply;
- Minimising the negative impact on the environment of energy use and production.

A programme 'Specific Actions for Vigorous Energy Efficiency' (SAVE) was adopted[14] by the Council of Ministers. A package of measures covered certifying CO_2 emissions due to energy use in buildings, thermal insulation of new buildings, regular inspection of boilers and cars, energy audits of businesses.

An Alternative Energy programme (ALTENER) to develop the use of renewable energy sources (REs), also adopted by the Council of Ministers[15], was intended to 'give impetus to RE development'. This comprised these features:-

- Promoting the internal market in REs;
- Financial and economic measures;
- Training, information and outreach services (such as promoting information centres as 'shop windows');
- Cooperation with other countries, particularly in the developing world, so that damage to forests and the world climate can be slowed down.

A further Commission proposal was a carbon/energy tax to apply pressure that would limit carbon dioxide emissions and improve energy efficiency. Half of the tax would be based on energy content and the other half would vary according to the carbon content. The tax would be fiscally neutral, since the proceeds would be used wholly for incentives for investing in energy-saving or CO_2-reducing measures. This was strongly opposed by the UK, disliked by other EU governments and remains for the time being still-born.

The Commission estimate that the EU accounts for 13% of world CO_2 emissions, with the USA contributing 23%, Japan 5%, and the East European countries and former USSR 25%. Even before the Rio Earth Summit, the Council of Ministers had decided in 1990 that global Community CO_2 emissions would be stabilised at their 1990 level by the year 2000 and the above proposals by the Commission were part of the strategy suggested.

UK Trends

In the UK[11], total energy consumption rose at an average rate of 2% a year to 1973 to the equivalent of over 200 million tonnes of oil (Mtoe) and after two periods of reduction related to Middle East wars regained that level in 1990. Through the recession it has stayed 'broadly stable'. But the composition of energy supply has markedly changed.

From the mid-1960's the UK has been using sharply increasing amounts of natural gas and also of what is called primary electricity — that is electricity mainly from nuclear sources, with a contribution from hydro. In earlier decades the use of petroleum has fluctuated but in recent years it has grown — mainly through the demands of transport — increasing at an average of 1½% per year since 1983. There was a sharp decrease in use of coal in the period of the coal strike of 1984, most of this fall being recovered in the following year but overall, the consumption of coal has reduced at an average of 2% per year since 1960.

The pattern of production has also undergone drastic changes. From 1975 petroleum production grew fast. The combined effect of the increases in supply of gas and petroleum was to transform the UK from a heavy importer of energy sources — in the early 1970s we depended on imports for over 50% of energy consumption — to a net exporter from 1980 for several years. But from 1988, the UK again became a net importer of energy, albeit on a relatively small scale, until 1993 when it achieved balance[16].

The magnitude of the revolution in energy supply can be gauged from the fact that in 1965 coal accounted for 96% of total energy production. In 1993, the last full year for which there are official statistics[16], coal supplied 18% of the total, natural gas 26%, petroleum 47%, with nuclear and hydro together 8%.

In UK official statistics the index applied to measure the relationship between energy consumption and economic activity is essentially the same as that used by the WEC — under their title of energy intensity — but compares the temperature corrected primary energy consumption with GDP at constant prices and calls it the 'energy ratio'.

As mentioned earlier in general terms, this ratio tends to fall, not every year, but certainly as a trend. From 1950 to 1973 it fell at an average rate of 1% pa for 23 years, then by about 2% pa until 1989. Since the middle of 1990 there has been continuing demand for energy in transport and in the form of electricity though the GDP actually declined for two years. So in the last few years, this ratio for the UK has risen.

Does the UK government have a policy for energy? In the report on 'Sustainable Development', the government states its policy to have the aim of ensuring secure, diverse and sustainable supplies of energy in forms that people and businesses want by means of competitive energy markets. 'Accordingly, the Government does not produce central plans' though it 'will continue to monitor the operation of the market'.

Although the government surprisingly closed down the Department of Energy as a separate ministry and incorporated its functions into the Department of Trade and Industry (DTI) it maintains relevant research. This includes a Coal Research & Development Programme, an Energy Efficiency Office and a range of research 'to improve understanding of climate change', an important part of this being the Hadley Centre for Climate Change Prediction and Research.

On new and renewable energy technologies, policy is to stimulate the development where they have prospects of being economically attractive and environmentally acceptable. It intends to 'work towards a figure of 1500 MW of new renewable electricity generating capacity in the UK by the year 2000'. There is a Non-Fossil Fuel Obligation levied on the Regional Electricity Boards and therefore in turn on the electricity user; 6% of this goes to supporting renewables and the balance to the nuclear industry. DTI also has a 'programme of research, development, demonstration and dissemination on new and renewable energy'.

What is the Pattern of this Review?

The following chapters will deal with individual fuels and with the renewable sources. Finally we turn to 'scenarios', 'energy Cases' and other forms of trying to assess the essentially unknowable future, that can form the basis for developing energy policies.

Footnotes

Definitions

OECD is the Organisation for Economic Co-operation and Development. Its members are:

Europe: Austria, Belgium, Denmark, Finland, France, Germany, Greece, Iceland, Republic of Ireland, Italy, Netherlands, Norway, Portugal, Spain, Sweden, Switzerland, Turkey, United Kingdom.

Other member countries: Australia, Canada, Japan, New Zealand, USA

LDCs in *BP Statistical Review*

Latin America, Africa, Middle East and Non-OECD Asia.

References in Main Text

1 *BP Statistical Review of World Energy* London: British Petroleum Co plc., 1993

2 *Report of the Intergovernmental Negotiating Committee for a Framework Convention on Climate Change on the work of the second part of its fifth session, held at New York from 30 April to 9 May 1992*

3 *The greenhouse effect: the scientific basis for policy* (a submission to the House of Lords Select Committee) London: Royal Society, July 1989

4 J Houghton. 'The predictability of weather and climate' *Philosophical Transactions of the Royal Society of London A* 337 (1991) pp.521-572

5 J T Houghton, G J Jenkins, and J J Ephraums (eds) *Climate change: the IPCC scientific assessment* Cambridge: Cambridge University Press, 1990. And *Climate change 1992: the supplementary report to the IPCC scientific assessment* Cambridge: Cambridge University Press, 1992

6 *Sustainable development. The UK strategy* London: HMSO, Jan 1994 Cmnd. 2426

7 *Energy-related carbon emissions: possible future scenarios in the United Kingdom* London: HMSO, 1992 Energy Paper no. 59

8 *Energy for tomorrow's world* London: Kogan Page for the World Energy Council, 1993

9 *Population summit of the world's scientific academies, New Delhi, India 1993* published as *Population - the complex reality* London: Royal Society, 1994 ISBN 0854034846

10 *Energy efficiency and conservation in the developing world* Washington, DC, USA: World Bank, Jan 1993

11 *Digest of United Kingdom Energy Statistics 1993* London: HMSO, 1993

12 *Basic statistics of the Community* 30th ed., Luxembourg: Eurostat, 1993 and later information from Eurostat

13 *A view to the future* European Commission, 1992 Special issue of *Energy in Europe* ISBN 9282636658

14 *Official Journal of the European Communities* (published Luxembourg: Office for Official Publications of the European Communities) 1991 p.L307/91

15 *Official Journal of the European Communities* 1993 p.L235/93

16 *Digest of United Kingdom Energy Statistics 1994* London: HMSO, 1994

Chapter 2

Coal

Contents

- Coal: Origins and Composition
  - Chemical Composition
  - Classification
- Sources and Reserves
- Using Coal and Protecting the Environment
- Clean Coal Technologies
  - Coal Extraction
  - Coal Preparation
  - Coal Combustion
  - Combined Heat and Power
  - Fluidised Bed Combustion
  - Coal into Gas
  - Underground Gasification
  - Hybrid Gasification and Other Clean Technologies
- Coal in the UK
- References

Coal

"This island is made mainly of coal...Only an organising genius could produce a shortage of coal"

Aneurin Bevan. Speech reported in *Daily Herald* 25 May 1945.

During the period when coal in Britain was really king, the record output was 287 million tons in 1913; of this 73 million tons were exported and 21 million supplied 'for the use of steamers engaged in the foreign trade' in the words of the old Department of Mines. Including the coal equivalent of the coke and manufactured fuel, the grand total shipped overseas amounted to 98 million tons[1]. These dazzling achievements were never repeated. In 1992/93 British mines produced a little over 81 million tonnes; almost 20 million tonnes were imported and (for the year 1992) a total of less than 1 million tonnes were exported[2]. Later in this chapter we discuss these remarkable changes. But first let us consider what is coal.

Coal: Origins and Composition

Though the fuel is now far less familiar in the UK and Western Europe than it was a generation ago, coal is still well-known as a blackish rock hewn out of the depths of the earth or — much more easily won — sometimes available near the surface. Rich in carbon compounds, this rock can be burnt. It is not a single mineral but a range of natural solids varying in composition. The complete range is broadly taken to extend from peat, through brown coal, lignite and bituminous coal to anthracite.

Coal is a fossilized product resulting from decomposition of abundant tropical forest plants under marshy conditions. Within it is a proportion of material that will not burn, such as sandstone and shale. This is largely the highly compressed residue of sand and silt deposited between and on the rotting vegetable matter. When coal is burnt, this incombustible material forms the ash.

In the swamps, plants grew thickly over long periods, then the area would subside and flood, ending the growth and resulting in the vegetable remains being covered by the sand and silt. Later the area was lifted by further earth movements and further swamp forests would grow. The process was repeated many times during periods measured in millions of years. Through a number of stages of biological, chemical and physical action the plants were chemically converted. The process is called 'coalification'; how far it has gone determines the 'rank' of the coal.

Generally, the process is more advanced in older seams, though coals are found of different rank even at the same geological age, since earth movements, pressures and temperatures have all varied greatly sometimes even in limited areas.

The chief coal-forming Period was that known as the Carboniferous, about 360 million to 280 million years ago. There are also some important deposits of lower rank dating from the Cretaceous Period, about 145 to 65 million years ago. The Coal Research organization of the International Energy Agency (IEA) now reports[3] a tendency for production to switch in recent years to the post-Carboniferous coal deposits of Cretaceous and other Periods formed under a wider range of environmental conditions. Peat is geologically recent, the product of less than one million years.

Chemical Composition

From the above, it is evident that the chemical composition of coal can be given only as ranges of values. They are shown for the main range of common coals in the table below from a US source. The moisture content is given for the coal as found; values for the other properties are given for the organic coal substance free of the associated moisture and mineral matter: -

Compositions of main range of common coals

Coal Type	Moisture %	Carbon %	Hydrogen %	Volatile Matter %	Calorific value, MJ/Kg
	As found				
Peat	90 - 70	45 - 60	6.8 - 3.5	75 - 45	17 - 22
Brown coals and lignites	50 - 30	60 - 75	5.5 - 4.5	60 - 45	28 - 30
Bituminous coals	20 - 1	75 - 92	5.6 - 4.0	50 - 11	29 - 37
Anthracites	1.5 - 3.5	92 - 95	4.0 -2.9	10 - 3.5	36 - 37

The so-called 'volatile matter' figure is the percentage weight loss when the coal is heated in the absence of air under standard conditions; it is therefore the result of decomposition. In the UK classification system5 it is used in conjunction with the coke type obtained under standard test conditions to measure the rank, or degree of coalification, of a coal. Test methods5 are of course standardised nationally and often internationally.

Classification

Coal may be classified in systems intended to serve scientific and/or commercial needs. There are both national and international classification systems. In an excellent survey[6] of the complexities of these systems, IEA Coal Research has examined their features and limitations. There is interest in classifying coals in order to enable a user to identify the most appropriate coal for a particular purpose. In addition the system should: —

- categorise coals;
- be able to allow assessment of any coal and be universally applicable to all coal types;
- allow an estimate or prediction of other properties or behaviour of the coal;
- have scientifically valid parameters;
- be simple to use;
- apply to single coals (though this issue is under debate, since coals may be blended for some applications);
- include both raw and washed coals;
- allow evaluation of potential environmental problems.

Since the traditional classifications were developed there has been increasing concern about environmental consequences. Consequently, IEA Coal Research has suggested that in a modern system, it would be useful to have parameters (such as one for sulphur) that would enable a user to evaluate potential problems in this area. Though sulphur and other constituents are often quoted in a coal specification, they have not in the past been part of the coal classification.

A scientific classification by a British coal chemist called Seyler in 1924 was based on plotting a graph of the carbon content of coals against their hydrogen content.

This produced a curved band corresponding to their degree of coalification. Other properties of coals, such as volatile matter and calorific value were added later forming a chart from which many properties of a given coal could be read off if for example, two properties were known[19].

The USA, Australia and Germany also developed their own systems. An international system[7] was proposed by the Economic Commission for Europe (ECE) in 1956. This has been replaced by an updated codification[8] intended to assist in the international trade in coal. In the codification eight basic parameters define the main properties of coal and are represented by a 14 digit code number. The number includes information about ash, sulphur and calorific value in the code. The International Organization for Standardization (ISO) is also working on a system, but this has not yet been published.

For brown coals, defined as those with a gross calorific value of less than 5,700 kcal/kg or 23.86 MJ/kg (calculated to a moisture and ash free basis) both the ECE and ISO have published classifications[9],[10]. This calorific value limit differs from that quoted in the table earlier and the ECE classification allows coals with a higher calorific value to be included if the coals are regarded in the country of origin as brown coals. The ECE also considers the boundary between peat and brown coal to be of no major economic importance.

Sources and Reserves

'Coal occurs more abundantly, and over a more extensive geographical area, than any other fossil fuel' points out the IEA[11]. Coal has been found on every continent in more than a hundred countries, and in sixty of these countries it is estimated to occur in quantities that are or could be significant. Proven economically recoverable reserves, are estimated to be enough for over 200 years' use at current rates of consumption, and are very much greater than those for oil or natural gas. In some countries coal is the only indigenous native fossil fuel, and even when it is imported, it is considered a more secure source of energy than other fuels because of the much wider range of possible suppliers.

Coal accounts for almost 30% of all the energy used worldwide. It is mainly used in power stations and these stations generate almost 40% of the world's electricity. Other major uses are in manufacturing steel and cement. Most coal is used in the region where it is produced and it is estimated that more than 60% of that used for generating power needs to be transported less than 50 km. Where it is important to minimise dust and noise on congested sites, coal can be transported pneumatically through pipes to enclosed silos. Less than 10% of production is sold between countries. Nevertheless, international trade in seaborne steam coal (that is coal intended for raising steam) is increasing — from 22 million tonnes in 1973 to almost 200 million tonnes in 1991 — and expected to increase further. The main net exporter is North America and the main net importer OECD Europe, but this is also changing with Asian demand expanding dramatically.

For 1992, the BP Statistical Review[12] gives world production of hard coal (bituminous coal and anthracite) as 3178.8 million tonnes, and that of lignite and brown coal as 1308.1 million tonnes. China was responsible for a quarter of total world production, with the USA producing a further quarter. The other big producer is the group of countries of the former Soviet Union, which account for 13.2%. Largest increases came from Asia and Australasia. For instance, Indonesia is beginning to be a major world producer, with Korea, Taiwan and Thailand as, what some call 'growing players in the world coal scene'. The ECE forecasts a continuing increase of world hard coal output of over 1% p a between 1991 and 2010. This corresponds with the trends we noted in Chapter 1, of a slow increase in coal use worldwide though the use of coal has been declining in the European Union and this decline is expected to continue.

Using Coal and Protecting the Environment

Growing emphasis on protecting the environment has resulted in ever greater pressure, from both public opinion and new laws, for minimising pollution due to emissions from processes. Many countries have now set their own national limits on emissions due to using coal, covering dust, sulphur dioxide (SO_2) and oxides of nitrogen (NO_x). In addition, at the Rio conference, developed countries also committed themselves to prevent increases in emitting CO_2. Coal is estimated to contribute almost one fifth to the man-made part of the greenhouse effect.

Contributions to the Enhanced Greenhouse Effect

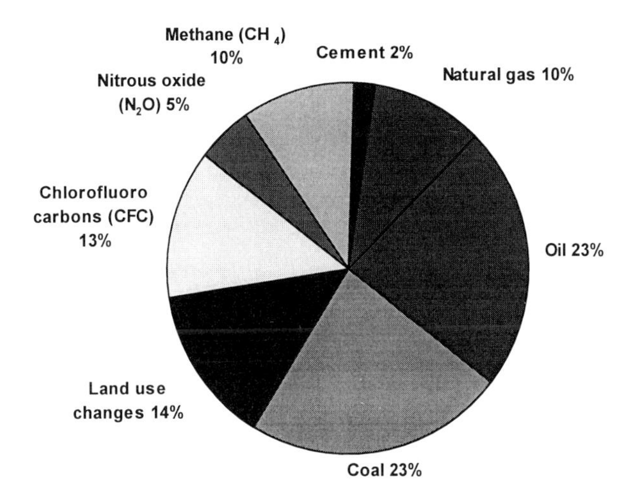

Source: Clean Coal Technology. Options for the Future. DTI 1993

A further problem is that when coal is burnt it leaves a residue of ash that also has to be dealt with in a socially acceptable way. Ash and the products of flue gas desulphurisation (FGD) are often tipped, under control, as waste landfill. Fly-ash, resulting from burning pulverised fuel, may be used for necessary filling as in building roads or foundations, and it is widely used in making concrete or building blocks. Gypsum produced in FGD systems using wet lime can be applied in making plasterboard.

Since coal is the fossil fuel with the highest carbon content, it is also the one that yields the largest quantity of CO_2 per unit of energy. Consequently the industry tends to focus on ways of dealing with this key environmental issue. Steady increases in thermal efficiency have been reducing the amount of coal burnt, and therefore CO_2 produced, per unit of electricity generated, or per unit of output in other applications. However, the industry recognises that 'if coal is to remain the backbone of energy supply security....ever more efficient technologies must be developed to enable coal to be used in an environmentally acceptable way'[3].

A review of the worldwide emissions to the atmosphere of methane — a greenhouse gas 21 times more potent weight for weight than CO_2 — has led to the comforting conclusion that the coal industry is but a minor contributor. The methane is evolved mainly during the mining of coal but methane emissions from natural causes and from the activities of the oil and gas industries were found to be far more significant.

Public concern and laws about controlling the discharge of SO_2 have led to both the use of coals with lower sulphur content and fitting of FGD systems to coal-fired power stations. A world survey by IEA Coal Research found FGD equipment installed in 18 countries covering 168 GigaWatt (GWe, where a GigaWatt is 10^9 Watt) of generating plant, and planned or under construction in nine other countries (over 107 GWe). When IEA Coal Research responded to further anxieties about bio-accumulation of elements such as arsenic, cadmium and mercury that occur in traces in coal, there was again a broadly comforting conclusion. Pollution control systems fitted to control emissions of particles and SO_2 also remove some trace elements. Some of the trace elements are attached to the solid particles and therefore trapped by electrostatic precipitators or fabric filters. Equipment to remove sulphur gases based on wet scrubbing also removes a large proportion of the more volatile trace elements.

A striking example[3] of a combined heat and electric power station designed to have minimum impact on the environment is the Avedoere Station in Copenhagen, Denmark. It comprises an enclosed coal handling system, noise suppression and emission control equipment; the building design won an open architectural competition.

Clean Coal Technologies

To combat the old stereotype of coal as a dirty and polluting fuel, the industry internationally has been promoting the concept of 'clean coal technologies' seeing them as strong options for the future[3]. These are technologies that will improve both the efficiency and the environmental acceptability of coal extraction, preparation and use. Though the greatest environmental impact arises in improving the way coal is used, there have also been big advances in the earlier stages of coal handling.

Coal Extraction

Underground mining, which wordwide produces about two thirds of the hard coal, has been ever more mechanised and automated. Modern guidance systems fitted to coal-cutting machines minimise the cutting of unwanted mineral matter. Health and safety of miners has greatly improved due to better hazard detection equipment, very low dust levels at the face, detection methods for methane and draining away the methane in appropriate cases.

In both surface and underground operations, the trends in many countries are for new projects to be subject to analysis and control of environmental impact. This will include effective management of waste and rehabilitation of the site. Many countries including the UK can show examples of sites greatly improved after rehabilitation and used for farming, building or recreation.

Coal Preparation

Even with the advantage of guidance systems on the cutting machines, the coal is delivered from the mine as a rather variable product containing mineral matter and undesirable elements, some of them in the mineral matter. Customers are tending to look for a consistent product — in size and composition — with little incombustible matter and may also call for low levels of sulphur, nitrogen and chlorine. So coal preparation may mean crushing, sizing, 'washing' (separating by the differing density of coal and dirt), drying or dewatering, and blending. All these traditional operations have been greatly improved by advances in instrumentation and control including continuous analysers of coal in motion. Coal cleaned to high standards could encourage new applications —- one surprising and impressive use being experimental firing in internal combustion engines[11].

Coal Combustion

Though coal can be chemically converted to other products, notably in three large-scale oil-from-coal plants by the petrochemicals and synthetic fuels group Sasol in South Africa[23], most coal is directly burnt. In power stations, the usual process is pulverised fuel (pf) combustion. This is based on grinding the coal to a fine powder and blowing it in with air to a large boiler. Almost all the carbon is burnt, but the ash forms dust that is largely captured usually by charging the particles and electrically trapping them in electrostatic precipitators. Alternatively the dust is caught in bag filters.

To minimise the amount of SO_2 emitted to the atmosphere, the coal may be treated in advance to reduce sulphur content, or combustion may be carried out in a way that traps the sulphur, or the flue gas may be treated in the process mentioned earlier known as flue gas desulphurisation (FGD). This process has variants, but the most usual form is wet 'scrubbing' (washing out in a tower) with lime or limestone and water, yielding a mixture readily converted to usable gypsum.

Power stations are now taking measures to reduce the formation of oxides of nitrogen (NOx) by using specially designed low NOx burners and/or a design known as 'staging'. In this design, air and fuel are introduced at different levels in the boiler furnace.

There are also processes for removing NOx from flue gases, where combustion control methods do not prove adequate. These further processes may use catalytic reduction or non-catalytic reduction. Combined processes for removing both SO_2 and NOx are largely under development but a plant known as SNOX (de-SOx and de-NOx) has been built at Vendysysselvaerket power station, Aalborg, Denmark.

Combined Heat and Power

Technical research has steadily increased the efficiencies of power stations but there are fundamental limitations to what can be achieved in a machine that converts heat into power. Heat has to be rejected from the plant at a relatively low temperature; this is known as 'low-grade heat'. Yet this heat can still be used for industrial or district heating so that the overall efficiency is increased. Such schemes are known as Combined Heat and Power (CHP), or Cogeneration schemes.

These schemes need suitable users for the heat reasonably close to the power station. For the best results the demand for heat and that for power need to be broadly 'in phase' and this has meant that the most favourable situations have tended to be industrial estates, although there are also many district heating schemes.

Fluidised Bed Combustion

Solids have some degree of rigidity, but broken down into particles they can be made to behave rather like a liquid by lifting and agitating them in a stream of liquid or gas. The process, known for about 60 years, is called 'fluidisation' and can be applied with great benefit to the burning of coal, by continuously feeding in coal to

Cogeneration Using Steam Extraction

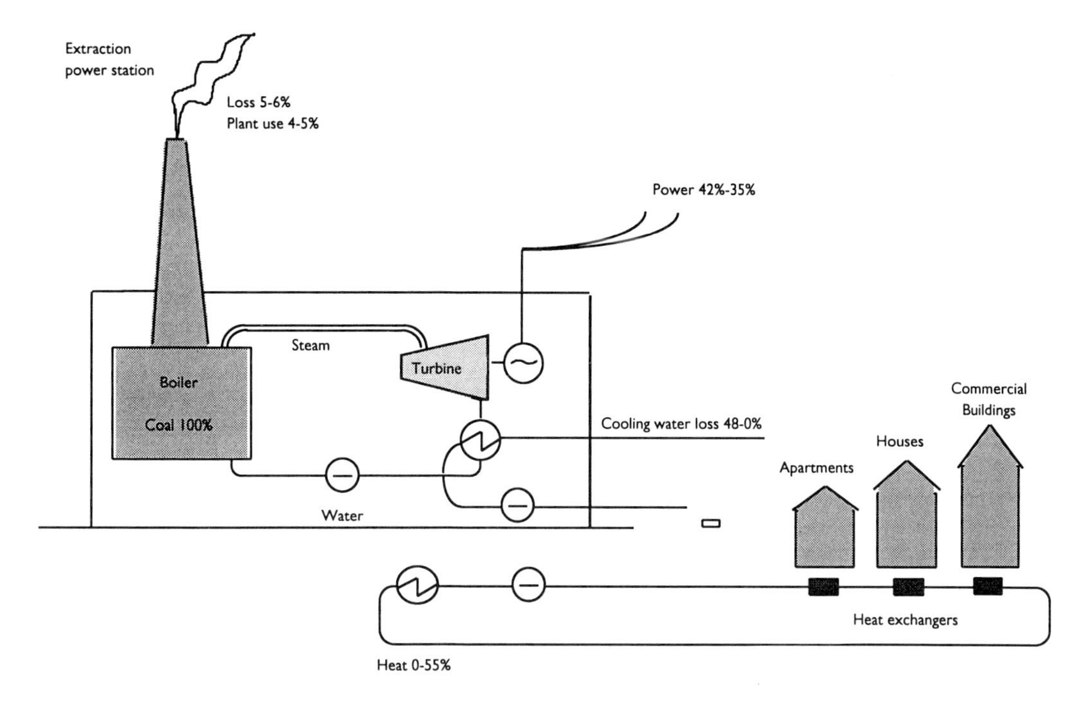

Source: Clean Coal Technology. Options for the Future. DTI for IEA 1993, p13
© OECD/IEA, 1993

the equipment and then tapping off surplus ash. The great advantages of burning coal on a fluidised bed include: —

- flexibility in taking a wide variety of fuels including those with high levels of mineral matter;
- very rapid transfer of heat to immersed water tubes;
- relatively low temperature of combustion and easy adaptation to air staging so that lower amounts of NOx are formed;
- facility for adding crushed limestone that can trap more than 90% of SO_2 formed during combustion;
- facility for operating to generate hot gas for gas turbines.

IEA states that there are more than 200 units working at atmospheric pressure in the industrial countries and a further 2,500 in China and India together. So far they have been demonstrated in sizes up to 160 MWe and a 250 MWe unit is due to be operating in the US shortly. More complex variants known as Circulating Fluidised Bed Combustors are also in use fired by a variety of fuels including chicken litter.

Pressurised versions of fluidised bed combustion have long been advocated mainly because the units can be made much smaller for a given duty. They can also be designed for operating in 'combined cycle' systems. In these the pressurised fluid bed burns coal that raises steam which is passed through a steam turbo-alternator and generates electricity. The gases from the combustion unit are under pressure and after cleaning can be passed through a gas turbine to generate further electricity. There are commercial units in Sweden, Spain and USA, with designs under development for units producing 340 Mwe.

Coal Into Gas

In the past, coal was used to produce gas by 'carbonizing' it, that is heating in the abscence of air, and simultaneously producing coke, coal tar and ammoniacal liquor. In the UK this process for making gas finally died away in the mid-1970's as natural gas started to flow into the country in substantial quantities, though some carbonization continues in order to make coke.

Modern gasification processes are based on coal being reacted with steam and oxygen at high temperature yielding a mixture of carbon monoxide and hydrogen. Heat is supplied by burning a small proportion of the coal. Systems employed may be based on a slowly descending bed of coal, or an adaptation of the fluidised bed bed, or on finely ground coal flowing with the steam and oxygen in what is called 'entrained flow'. The gas produced is cleaned and may be used for fuel or for chemical synthesis.

If the gas is used as a fuel, an attractive option is to apply it in combined cycle systems. An adaptation of a natural gas fired process is an integrated coal gasification and combined cycle system (mercifully known as IGCC). In this form of combined cycle hot gas from the gasifier, which is under pressure, is cleaned and fed to a gas turbine to generate power. Waste heat from this turbine, with other recovered heat raises steam that generates further electricity in a turbo-alternator. Nett efficiency is reported to be around 43%, which is higher than that of the most technologically

Errata

We would like to apologise to readers for the following errors which came to light too late to be corrected:

Acknowledgements
The NRPB is the National **Radiological** Protection Board

Page 11, paragraph 3. The first two sentences of this paragraph should read: "Proven reserves of gas increased about fourfold between 1967 and 1992. In the latter year, the main growth of reserves was in the former Soviet Union and the Middle East."

Page 21, table. The words "on dry, mineral matter free basis" should run across the table under the headings "carbon, hydrogen, volatile matter, calorific value".

Page 32, paragraph 1 The correct depth is 550m.

Page 34, last word. "It" should read "Its".

Page 49, table. In box headed "Feedstocks" "Vacum" should read "Vacuum".

Page 84, paragraph 1. The number of the reference should read 31.

Page 85, last word of last paragraph before table. Reference 32 has been omitted.

Page 89 Reference 29. The word "offer" should read "Offer" (the Office of Electricity Regulation)

Please also note the following missing references:
31 *Information on the Non-Fossil Fuel Obligation for Generators of Electricity from Renewable Energy Sources* (*Renewable Energy Bulletin 5* (Department of Trade and Industry) (October 1993)), p9.

32 *Radiation. Doses, Effects, Risks* 2nd edition Oxford: Blackwell for United Nations Environment Programme, 1991. ISBN 0631183175.

Page 119, paragraph 1, line 1. "thst" should read "that".

Page 135, table. Column headed A should appear under "2020 Calculated".
"Ccalculated" should read "Calculated".

Page 146, table.

C3 The phrase beginning "cyclist deaths..." should read "cyclist deaths from 4.1 per 100 million kilometres to 2 per 100 million kilometres".

D1 The phrase beginning "public transport..." should read "public transport from 12% in 1993 to 20% by 2005 and 30% by 2020."

E line 2 "lthe" should read "the".

F1 "emmissions" should read "emissions".

F3 "dutty" should read "duty".

Page 147, paragraph 2. "Meterorological" should read "Meteorological".

Page 167. Ozone Layer. "O^3" should read "O_3".

Throughout. "Naptha" should read "Naphtha".

The British Library
Science Reference and Information Service

Integrated Gasification Combined Cycle Power Generation

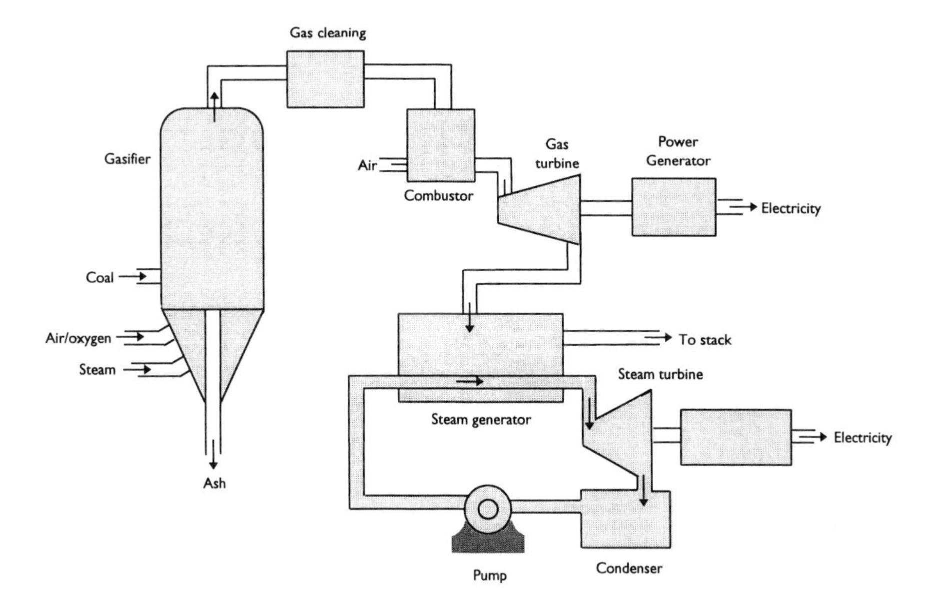

Source: Clean Coal Technology. Options for the Future. DTI for IEA
© OECD/IEA, 1993

advanced pulverised fuel plants. This system also gives lower levels of emissions of SO_2, NO_X, and CO_2. Demonstration plants are planned in several countries and the petroleum industry has regarded these developments as serious enough to justify a special conference of the Institute of Petroleum to assess how soon this group of technologies will be a threat to the oil and gas industry.

Underground Gasification

For over a hundred years there has been interest in the concept of gasifying coal as it lies in the ground. Holes are driven down to the coal, the coal fired and supplied with steam and oxygen, and gas is drawn off. This approach is particularly attractive for deriving useful energy from seams that are unworkable by normal means. Unfortunately the process has proved difficult to control and the gas is of low and variable quality. In 1976 a report[13] by the National Coal Board concluded: "...at present the costs lie at the upper end of energy costs and would have to be reduced if the process were to be operated on a commercial scale...It merits continued observation to monitor changing economic relationship between this and other forms of energy supply." However, Spain, Belgium and the UK, with EU support are currently running a field trial of underground coal gasification in the Teruel region of Spain on a seam at a depth of 550mm. If this is successful, it is hoped to develop techniques to access the very large coal reserves available at depths up to 1000m in Western Europe.

Hybrid Combined Cycles and Other Clean Technologies

Hybrid systems include two-stage processes based purely on coal. The fuel may be partly gasified yielding a fuel gas to be fed to a high temperature gas turbine cycle and leaving a solid residue known as 'char'. This char is burnt in a separate combustion unit to produce steam for generating electricity. The version of this developed by British Coal at its Coal Research Establishment, called the Topping Cycle, was assessed by a DTI Working Party in 1992 as more cost-effective than competing IGCC systems. But an experimental facility at Grimethorpe was nevertheless closed down a year a year later awaiting further funding. There are also other hybrid designs including combinations of natural gas and coal combustion used in a working plant in Germany.

The IEA review of clean coal technologies[11] also refers briefly to these further options. Coal may be converted to liquid fuels (petrol and diesel fuel) by gasifying it and using the gases produced for chemical synthesis or by direct methods of conversion applying liquid solvents to the solid coal. There are projects for coal-fuelled heat engines in the form of direct fuelled systems, and indirect gas turbine systems. Coal-water fuel has been successfully used in locomotive and industrial diesel engines. In devices known as fuel cells, fuel is directly converted into electricity without using a heat engine. Current units are generally supplied with natural gas but there are proposals for using fuel from coal gasifiers. A further direct system called 'magnetohydrodynamics' (MHD) is also thought to hold promise for the future. It is essentially a two-stage process. Coal is burnt to produce an extremely hot gas and a chemical compound added as a 'seed'. In the electrical stage, this gas is passed through a strong magnetic field and electricity is tapped from the gas stream.

Can CO_2 Be Trapped or Used?

Since so much importance is attached to the effects of CO_2 on climate, IEA has established a Greenhouse Gas R&D Programme as a collaborative project of some 14 countries and the EU Commission. Going beyond the improvements in efficiency of using fuel discussed earlier, this programme is also examining such counter measures as natural fixation of CO_2 (as in afforestation), disposal methods (possibly in the deep ocean or in saltdomes), or recycling in industrial processes.

Coal in the UK

When coal was nationalised in 1947, the nation took over more than 1400 collieries. Of these 480 were small mines with fewer than 30 men underground; they were granted licences to continue in private operation and the remainder - in the words of the official notices - were 'managed by the National Coal Board on behalf of the people'. At that stage, coal supplied 93% of UK energy demand and the emphasis was on expanding coal production to prevent shortages of supply.

By the late 1950s and early 1960s the pattern had changed and demand for coal was falling. Then, in a further twist of the energy market, OPEC quadrupled oil prices during 1973-1974.

There had been national coal strikes in 1972 and early in 1974, but after a change to a Labour government in 1974, a tripartite group of government, National Coal Board (NCB) and mining unions accepted a new Plan for Coal based on increased government financial assistance, closing older pits, 'creating 42 million tons of net colliery capacity to offset inevitable exhaustions', and increasing opencast production. The fall in coal demand was indeed stayed for a period but then resumed[20]. In 1979 a Conservative government was returned to office. Within a few years a combination of factors - economic, political and social - led to a major coal strike, the flash point being a colliery closure announced without the normal review procedure[21].

In 1984 before this national strike, there were 170 pits with 190,000 miners. By the early part of 1994 on the run-up to privatisation there were 17 collieries with 12,000 men. A series of sociological studies[22] by university researchers and others from several countries have revealed the partiality of actions by state agencies during the strike, and the social and economic deprivation that has resulted in mining communities as a consequence of the strike.

Consumption of coal has fallen in the UK, irregularly since 1913, and fairly steadily since the early 1960s as users turned to fuel they found more convenient - oil when it was cheap, gas and electricity. The only major users increasing their use of coal were for many years the electricity generators, and more recently they have had a large contribution from nuclear sources. Overall the effect has been that total inland consumption of coal was halved[2] between 1960 and 1992.

In 1992, electricity generation accounted for almost four-fifths of all coal consumption. However, since the electricity supply industry has been privatised, there has been the famous 'dash for gas'. The generators planned contracts to buy 40 million tonnes of coal from British Coal in 1993/94 and up to 30 million tonnes in each of the following years until 1997/98.

Announcement of plans by British Coal, endorsed by the government, in 1992 to close 31 mines led to widespread uproar and demonstrations in the country. Earlier there had been a report by a Commons Select Committee[14, 15] urging some degree of government assistance to the industry. After further consultations, the government issued a White Paper[16] and a summary of the findings[17]. This offered temporary support to the industry but otherwise left it open to a competitive energy market. There was also a limited package of support measures for areas affected by closures (though in practice, there continues to be heavy unemployment in former pit areas). The government stated that it planned to privatise the industry as soon as possible. Actually, British Coal while still in public ownership showed some remarkable improvements in productivity and in costs reduction; in the year from March 1992 overall productivity in deep mines increased by over 35%. Nevertheless, the Coal Industry Act was passed early in 1994 and privatisation of the industry was due to be complete by the end of that year. In autumn 1994 a new Government quango, known as the Coal Authority, took over ownership of the country's coal reserves. This Authority became the main licensing body for mining.

The government is also supporting research into new coal technologies. The Department of Trade and Industry has set up a Coal R&D programme of some 63 projects. At March 1993, it announced that the total contract value of its new projects was over £13 million, including a DTI contribution of £4.5 million. It reports[18] point out that clean coal R&D is an international activity; they refer to many links with the IEA research projects summarised above and specifically with USA and German projects.

The government Summary document concludes 'the coal industry should be able to ensure that coal continues to play a significant part in meeting Britain's energy requirements in the future'[17].

References

1 *Report on the British coal industry* Political and Economic Planning, 1936

2 *Digest of United Kingdom Energy Statistics 1993* London: HMSO, 1993

3. *Annual Report 1992 - 1993. IEA Coal Research* London: IEA Coal Research, 1993 (This lists a wide range of further detailed individual reports covering many of the subjects in this chapter)

4 Adapted from original table in 'Coal' article in *Kirk-Othmer Encyclopaedia of Chemical Technology* 4th ed., New York: Interscience, 1993. vol. 6

5 *The coal classification system used by the National Coal Board* (Revision of 1964). London. National Coal Board. Test and analysis specifications include:

BS 3323: 1992 *Glossary of coal terms* London: BSI, 1992

BS 1016 *Methods for analysis and testing of coal and coke* London: BSI (This comprises up to 25 parts, dated up to 1994, dealing with different constituents, calorific value, physical properties and methods of reporting results)

BS 1017 Part 1: 1989 *Sampling of coal* London: BSI, 1989

6 Anne M Carpenter *Coal classification* London: IEA Coal Research, Oct 1988 IEACR/12 ISBN 9290291621

7 *International classification of hard coals by types* New York, USA: United Nations, 1956 E/ECE/247, E/ECE/COAL/110

8 *International codification system for medium and high rank coals* New York, USA: United Nations, 1988 E/ECE/COAL/115

9 *International classification of brown coals* New York, USA: United Nations, 1957 E/ECE/297, E/ECE/COAL/124

10 *Classification of brown coals and lignites* Geneva: International Organization for Standardization, 1974 ISO 2950

11 *Clean coal technology. Options for the future* London: DTI for international Energy Agency, Paris, 1993

12 *BP Statistical Review of World Energy* London: British Petroleum Co plc., 1993

13 P N Thomas, J R Mann, F Williams *Underground gasification of coal* London: National Coal Board, 1976

14 *House of Commons Trade and Industry Committee: Session 1992/3, first report: British energy policy and the market for coal* ISBN 010223793X

15 *House of Commons Employment Committee: Session 1992/3, second report: Employment consequences of British Coal's proposed pit closures* ISBN 0102844933

16 *The prospect for coal: Conclusions of the Government's coal review* Cmnd., 2235 ISBN 0101220421

17 *The prospect for coal. Summary of the coal review White Paper* London: DTI, March 1993

18 *Coal Research and Development Programme. Progress report for the Department of Trade and Industry 1992-93* Harwell, Oxfordshire: ETSU, 1993 Report no. COAL R015 and further associated reports

19 C A Seyler *Fuel* vol. 3 (1924) pp.15, 41, 79; *Proceedings of the South Wales Institute of Engineering* vol. 53 (1938) pp.254, 396

20 *Coal for the future. Progress with 'Plan for Coal' and prospects to the year 2000* London: Department of Energy, 1977; and Israel Berkovitch *Coal on the switchback* London: George Allen and Unwin, 1977

21 John Lloyd *Understanding the Miner's Strike* London: Fabian Society, 1985; and John Chesshire *The future for coal* London: Fabian Society, 1984

22 For example Johnathan Winterton 'The end of a way of life: coal communities since the 1984-85 Miners' Strike' *Work, Employment and Society* vol. 7 (1 Mar 1993) pp.135-146; and Karlheinz Durr 'Der Bergarbeiterstreik in Grossbritannien' *Politische Vierteljahresschrift* (West Germany) 26(4) 1985 pp.400-422

23 *Annual Report 1993. SASOL* Johannesburg: SASOL, August 1993

Chapter 3

Petroleum

Contents

- Origins and Composition
- From Crude Oil to Useful Products
- World Patterns; Production, Consumption and Trade
- Environmental Effects
- Chemicals
- UK Trends
- References

Petroleum

"Everything is soothed by oil"

Pliny the Elder [Gaius Plinius Secundus]. *Natural History*, Book II, 234.

As noted earlier in Chapter 1 transport is one of the two fastest growing sectors of energy use (the other being electricity).

Transport overwhelmingly depends on oil products[1]; as do the chemical industries, a key group of producers in modern society.

The term 'oil' is applied to many liquids, usually viscous, flammable, insoluble in water but soluble in organic — that is carbon-based — solvents. So it includes fatty oils from animal, vegetable or marine sources, and essential oils, generally with characteristic odours, derived from some plants. But our main concern here is with petroleum — 'rock oil' — one of the group of mineral oils that also includes oils from coal and shale and consists largely of hydrocarbons, compounds based on carbon and hydrogen. In this chapter the term 'oil' will be used as a brief synonym for petroleum, by far the most important of the mineral oils.

Origins and Composition

Though seepages of crude oil are reputed to have been noted and sometimes used several thousand years ago, its modern history is taken to have originated with the discovery of petroleum by a man called Edwin L Drake in Titusville, Pa, USA in 1859. Drilling and development then rapidly spread in different parts of the Americas, then Romania and Iran[2]. Discoveries were made in Iraq in 1923, Bahrain in 1932, Saudi Arabia and Kuwait in 1938.

Petroleum geologists consider that petroleum derives from organic matter of both plant and animal origin that accumulates in fine-grained sediment under quiet conditions relatively deficient in oxygen. As the sediment becomes buried, the petroleum is generated under the combined influences of heat, pressure and time. The liquid is then squeezed out from the source rocks as they are compacted. Both the source rocks and the rocks through which it moves are thought to have catalytic effects (that is speeding up the chemical reactions).

Most of the world's oil is present between the depths of about 600 and 3000 metres, Because time is a major factor, it is impossible to try to copy the process rigorously in the laboratory. The geochemical arguments are based on the transformations of organic matter in the rocks as found in chemical investigations.

Petroleum liquid and gas move upward through permeable material until they reach either a surface or a rock structure that forms a trap. These traps are classified as anticlinal (arch-shaped), fault (due to a break in continuity of strata), and stratigraphic (due to the position of the strata).

In all cases there must be a porous and permeable reservoir rock, sealed above by a fine-grained relatively impermeable bed. The latter may be of clay, shale, marl or salt. Almost 90% of known petroleum reserves are in anticlinal or fault traps; the two together are called structural traps. Under the oil in the reservoir is water and there may be gas forming a top layer or cap.

The material as it occurs in nature is known as 'crude oil'. In the petroleum industry crude oil is distinguished from the natural gas liquids. These are relatively volatile liquid hydrocarbons and can be completely distilled — that is vaporized by heating, separated by differences in boiling point and then condensed. Crude oil consists of liquid mixtures that can flow or be made to flow up a well-pipe but cannot be completely distilled. The crudes contain non-volatile constituents, known as residual components or residue. Also excluded by this classification are the hydrocarbons that call for processing in order to extract them from the earth. That means shale oils that need strong heating to win them from shale rocks, the semi-solid bituminous tar sands of Athabasca in Canada, and pitch lakes as in Trinidad.

Exploration goes on continually and has resulted in more than doubling the proved reserves of oil over the last 25 years. Most of the major increases have been in the Middle East, which accounts for two thirds of all world proved reserves, with Saudi Arabia alone sitting on over a quarter of the world reserves[3]. Next in world importance comes Latin America with over 12%, most of this being shared between Venezuela and Mexico. In Africa — with a total of 6% of world reserves — the main deposits are in Libya and Nigeria, with a smaller proportion in Algeria. Britain's share, though the oil has meant so much to the national economy, is estimated to amount to only some 0.4% of the world proved reserves.

Chemically, crude oils consist mainly of hydrocarbons —compounds of hydrogen and carbon but they are complex mixtures, containing many members of several series of hydrocarbons and small proportions of compounds also containing the elements sulphur, nitrogen, vanadium and nickel. Density may be within a range of 0.780 up to 1.000 kg/litre at 15C. The True Boiling Point curve may extend from 50C to the equivalent of 550C at normal atmospheric pressure, leaving a residue. (This expression 'True Boiling Point', TBP, is used because the test distillation is carried out in standard conditions where the pressure is reduced in stages so that higher boiling components can be distilled at temperatures low enough to prevent them from decomposing. Then the temperatures are corrected to what would have been the equivalent at atmospheric pressure). The TBP curve is applied to estimate the yields of distillate and residue that will be obtained at particular temperatures from a crude oil in a commercial, continuous distillation plant. Several methods have been standardised for testing petroleum and its products, the British Standard covering this area[14] being published in multiple parts. They are generally identical with the corresponding test methods of the Institute of Petroleum, International Standards and European Standards.

From Crude Oil to Useful Products

Crude oil is distilled in refineries into a number of cuts or fractions of groups of compounds and these are further intensively processed by a battery of physical and chemical techniques. Generally, the smaller the molecule of the constituent compound, the lower its boiling point — or put alternatively, the more volatile it is. Though there are variations in procedures depending on the compositions of the particular crude, broadly the main fractions at this first stage, starting from the most volatile, tend to be these: —

- Petroleum gases, or refinery gases;
- Gasoline;
- Naphtha, largely used for chemicals;
- Kerosine or paraffin;
- Gasoil (so-called because it was originally used for enriching town gas) or diesel;
- Lubricating oil and wax;
- Bitumen.

Further processing yields a range of products — a range that can be flexibly varied to meet new or changing needs. These products can be grouped[2] in this way: —

- Those that can be made to explode under control in combination with air; they are used in prime movers such as the internal combustion engine in vehicles or in static engines;
- Those that burn providing heat and light; the main representatives are the gases, kerosines, gas oils and fuel oils;
- Hydrocarbons used as lubricants, asphalt, propellants for sprays, solvents, waxes;
- Non-hydrocarbons such as sulphur, vanadium, acid sludge; these may be waste or may justify recovery if they can find economic uses;
- Feedstocks for processing to gas, a wide variety of chemicals, and for growing protein.

Treatments applied in addition to distillation and blending include desulphurisation, the sulphur being sold to chemical companies, and 'cracking'. Chemically, this term is used to describe the breaking down of larger molecules into smaller ones — the effect being that viscous, high-boiling materials are converted into more mobile, more volatile, and most important, more valuable materials. Naturally, crude oils with a higher proportion of more volatile constituents, known as light or ultra-light crudes, are more valuable than heavier ones that call for more processing. Saudi Arabia has discovered large new reserves of light crudes and plans to step up their production. The report adds that refiners who have been investing heavily in cracking units to break down heavy crudes might have to reconfigure their plants to take lighter grades.

Cracking may be effected by heat alone (thermal cracking), or more usually with the aid of catalysts (catalytic cracking), or with hydrogen that chemically 'adds on' to the molecule (hydrocracking). Several different types of chemical reactions may be applied to increase the octane number (measuring the anti-knock properties of a fuel in an internal combustion engine) of constituents of petrols or motor gasolines. Among them are : —

- catalytic reforming — with the aid of a catalyst changing the material chemically into an appreciably different one; this is carried out with at least four different chemical reactions;

- isomerisation — changing the arrangement of the atoms within the molecule (though not their numbers), notably changing straight chain pentane and hexane to the branched chain compounds known as the iso-compounds, again using a catalyst;
- catalytic polymerisation — again speeded by a catalyst joining together molecules of the lower hydrocarbon gases into liquids that can be components of aviation gasoline giving high performance.

Lubricating oils also need further processing that may include further distillation, purifying with fuller's earth, solvent treatment, even hydrogenation, and blending.

World Patterns; Production, Consumption and Trade

Though the Middle East is the home of such a vast proportion of oil reserves, its production amounts to only 28% of the world total, Saudi Arabia contributing almost half of this, with Iran and Abu Dhabi as the other major producers[3]. In Africa, in descending order of output Nigeria, Libya, Algeria and Egypt are the largest producers, the total for that continent being about one tenth of the world production. Asia also contributes about a tenth, almost half of this coming from China, and a quarter of it from Indonesia. Latin America with one eighth of world output has several producing countries the most important being Mexico and Venezuela.

The remaining producing areas are North America (16.2% of the total), OECD Europe (7.2%), and Non-OECD Europe (14.6%), meaning largely the former Soviet Union. If the former Soviet Union is considered as one territory, then that is the world's largest oil producer, most of the production coming from Russia. Since 1987 the trends have been for production to fall in the USA and in Russia; but this has been more than offset by increasing output in other areas, notably in the Middle East where most of the increase came from restoring Kuwaiti facilities. In the former Soviet Union, there are reports[6] that the large number of joint ventures are leading to a metamorphosis. The region is being transformed in a way and at a speed which would not have been considered possible even a few years ago. 'Western oil and gas activity in the FSU has taken off'.

Turning to oil consumption, the most striking feature is the great difference in usage between the average 3 tonnes per head in North America in 1992 and the far lower averages elsewhere — about 1 1/2 tonnes for OECD Europe, 1 tonne for Non-OECD Europe, and 1/4 tonne for the rest of the world. In the USA gasoline remains the dominant product with a much smaller, and declining, proportion of fuel oil consumption. The less developed countries tend to use a larger proportion of fuel oil than gasoline but in these countries the use of both gasoline and what are called middle distillates are growing strongly.

Marked differences in the proportions produced and those consumed in different parts of the world are reflected in heavy world trade in oil[3]. The Middle East is by far the largest exporter of crude, shipping out over half of all world exports of crude and also about a fifth of world exports of products. The USA is the heaviest importing country (from Latin America and the Middle East), while as would be expected Japan and OECD Europe are the other major importers of crude and products.

Cut of the Demand Barrel 1992

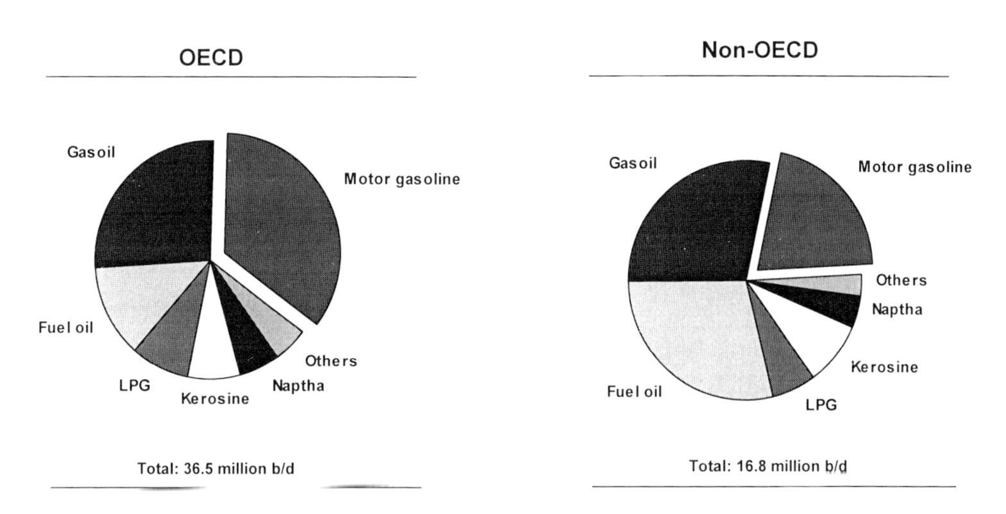

Source: Energy in Profile, Number 3 1993. Shell

Efficiency in using energy is discussed in a later chapter but we should mention here that there have been improvements in fuel economy in cars[11], most dramatically in the USA, the largest user and the one that most needed the improvements. Fleet averages for fuel use in new cars in the USA are reckoned to have fallen from 17 litres per 100 km in 1970 to 9 litres by 1989. For aircraft the average increase in fuel efficiency for the period 1973 to 1990 is estimated to have been 3.8% per year.

Prices of petroleum, traded worldwide and for many decades of such immense world importance, have naturally been influenced by relevant world events.

In many of the producing countries, notably in the Middle East, the earnings from petroleum have provided the main part of the countries' national income. After a complex series of events that reduced their revenue a number of them formed the Organisation of Petroleum Exporting Countries (OPEC) in 1960 in order to improve the bargaining strength of those countries relative to the oil companies'. BP's major handbook on the industry[2] summarises the consequences in this way; "...in 1973..effective ownership of production was transferred from operating companies to governments or their national oil companies and with it an end to control over the development, production and price of the bulk of the world's internationally traded oil by companies operating in the OPEC area". Early effects were indeed to drive up the price of oil, particularly after the so-called Yom Kippur war in the Middle East in 1973, and the associated use of the 'oil weapon' of cutting Middle East supplies.

However, since reaching a high point in the early 1980s, prices have fallen and in more recent years the industry has experienced an extended period of low prices which have 'taken [a] toll on oil companies and producers alike. The age of the tanker fleet is beginning to cause alarm'.[7] Increasing social pressure for protecting the environment has resulted in tighter laws governing the operation of refineries and demands for products of better quality. Recent trends are reported to have added greatly to the industry's costs, for example in building gasoline terminals. Responses by the industry have included widespread rationalisation and reducing its workforces. Improvements have also been made in 'refinery loading'[3]. Refinery throughput has tended to increase, albeit a little irregularly, while refinery capacity has tended to be cut back since 1980 (after growing rapidly through the previous twenty years).

Environmental Effects

Because fuel is not completely burnt in engines, transport worldwide is estimated to contribute well over 80% of the emissions of the poisonous gas carbon monoxide, and about half of the unburnt hydrocarbons. Engines also produce SO_2 and smoke, while the high temperatures within engines promote reactions that produce NOx from the air. All of these are objectionable and/or poisonous to humans; in contributing to acid rain they attack forests and buildings, contaminate soil and water. Activated by sunlight, NOx reacts with hydrocarbons and oxygen to form low-level ozone and the notorious photochemical smog.

Improvements in the efficiency of cars noted above have been accompanied by ever-tightening standards for reduced emissions from vehicles in the industrialised countries[11] and more are scheduled for the future. To meet these stricter standards, car designers and manufacturers have aimed to improve engine design and to introduce catalytic converters and activated carbon canisters in cars fuelled by gasoline. In mid-1994, the press reported[20] that the US Administration was entering into alliance with the motor companies to develop a low energy car.

Generation of CO_2, the principal greenhouse gas, per unit of distance travelled is naturally reduced per vehicle by the improvements in fuel efficiency but one method of increasing fuel economy and power output — that is by adding lead compounds to prevent 'knocking' — is now being phased out and alternative methods applied. The higher the compression ratio of an engine, that is the ratio of cylinder volume with the piston pushed out to the volume with the piston fully pushed in, the better its fuel economy. But increasing this ratio is limited by the start of 'pinking' or 'knocking'; this is the noise made when the fuel gases explode spontaneously before the flame in the cylinder reaches them. The anti-knock property is measured in terms of 'octane ratings'[19]. In the 1920's it was discovered that tetraethyl lead and tetramethyl lead greatly improved the octane quality. More recently, there has been great concern about the poisonous effects of the lead necessarily discharged into the atmosphere as a result, particularly when investigations showed that it reduced the IQ of children exposed to appreciable concentrations. The industrial countries have now set a succession of ever lower limits for permitted lead in gasoline. In turn, this has also led to changes in gasoline composition to raise octane level by other means, and in many cases changes in the valves of cars since the lead in practice acted as a valve lubricant. A further advantage is that cars using unleaded gasoline can be fitted with catalytic systems (otherwise poisoned by lead) for emission control, that will normally destroy unburnt hydrocarbons, oxidise the poisonous carbon monoxide to the dioxide and decompose NOx to its harmless constituents nitrogen and oxygen.

The effect on CO_2 generation of replacing petroleum fuels by alternative fuels when all the processes involved are taken into consideration is not clear-cut, and needs separate studies. Renewable sources of energy and alternative fuels will be discussed in a later chapter.

Yet despite the legal pressures to reduce pollution and the favourable trends per vehicle, there are both lags in the effects due to the slow rate of replacing older vehicles, and counter-effects due to increasing numbers of vehicles, increasing congestion in cities, and the ever-increasing sprawl of cities. In attempts to mitigate pollution in cities, public authorities have introduced central pedestrian precincts, and some have introduced road charging in city centres and improved public transport. These factors have been the subject of studies on city demographics and modes of transport[11] For instance petrol (gasoline) consumption per head is inversely proportional to the population density in the extended city. Petrol use per head is high in low density Houston and Phoenix in the US, intermediate in medium density European cities and lowest in densely packed Hong Kong. A further important factor is the growth of what are called 'edge cities' in broadly prosperous regions with high car ownership around big cities. They depend on high usage of cars.

Where petroleum products are used as industrial fuels, there are also major efforts to improve efficiency. It is likely that the main objective is to increase economy but the effects also include reducing CO_2 emission per unit of energy. This applies to combined heat and power schemes discussed earlier that can be based on any primary fuel including petroleum fuels. There are also major research efforts devoted to improving efficiencies of gas turbines fired by various fuels[13], including of course those from petroleum sources. Standardising properties and composition also contributes to efficiency in use, and petroleum fuels (and other petroleum products) are covered by a range of British and International Standards[15].

Even during its use in ships and its transport, oil is liable to cause liquid pollution of the sea, sometimes on a catastrophic scale when tankers have run aground. A succession of conferences on marine pollution have resulted in an International

Convention known as MARPOL 73/78, for preventing pollution from ships[16]. The International Maritime Organisation (IMO) estimate that 'in 1981 some 1,470,000 tons of oil entered the world's oceans as a result of shipping operations'. As a result of the impact of MARPOL, IMO believe that by 1989, this pollution was reduced to 568,800 tons.

IMO also expect this favourable impact to be even greater in the future. Many older tankers will be scrapped during the next few years. They will be replaced by new ships that will comply fully with the Convention and generally be built to higher standards. This will reduce the likelihood of both accidental pollution and fouling from operations such as discharging from oil tankers of machinery wastes or tank washings.

In addition, work greatly stimulated by the need to clean up the devastation caused by the Exxon Valdez disaster (Valdez, Alaska, 1989), showed the effectiveness of bioremediation[18] in clearing spilled oil. This is the use of microorganisms sprayed on to the oil to biodegrade it. The activity of the organisms is greatly enhanced by adding nutrients containing nitrogen and phosphorus. Using bioremediation, beaches in Alaska were cleaned up, both at and below the surface, at rates estimated to be between 3 and 15 times the rate for natural recovery.

Chemicals

Evidently a large part of producing the main products from petroleum is of the character of chemicals production. Coal, vegetable and animal materials were originally used in producing organic chemicals, but the pattern changed markedly in the 1940s and particularly in the 1950s. A range of synthetic rubbers, detergents and plastics were industrially manufactured in ever increasing amounts from petroleum sources. Today it is estimated[4] that about 90% of organic chemicals are derived from oil and natural gas. Producing the base chemicals needed as what are usually called the 'building blocks' for usable chemicals is closely integrated with primary oil refining. In 1992 about 9% of world oil was being used as chemical feedstock.

Naphtha is the feedstock most often used for chemicals in Western Europe and Japan, but the USA tends to use ethane and propane from natural gas liquids as the main feedstock. To think of these broad generalisations as a rigid division of use between the various petroleum products would be to underrate the flexibility of chemical 'manoeuvring' in applying these materials. On the one hand, most naphtha is processed to motor gasoline. On the other is the fact that base chemicals are obtained as by-products in other operations including the processing of naphtha to gasoline. Chemical feedstocks are also obtained from natural gas liquids, refinery gases, gasoil, petroleum wax and fuel oil. Decisions on feedstocks and their uses are determined by considering interacting factors of prices, availability, specific demands and plant available.

Yields of Ethylene and Co-Products From Various Feedstocks

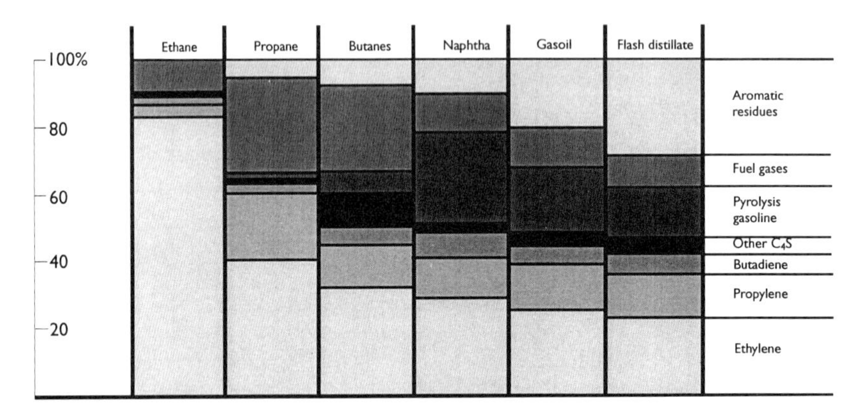

Feedstocks for Ethylene Crackers

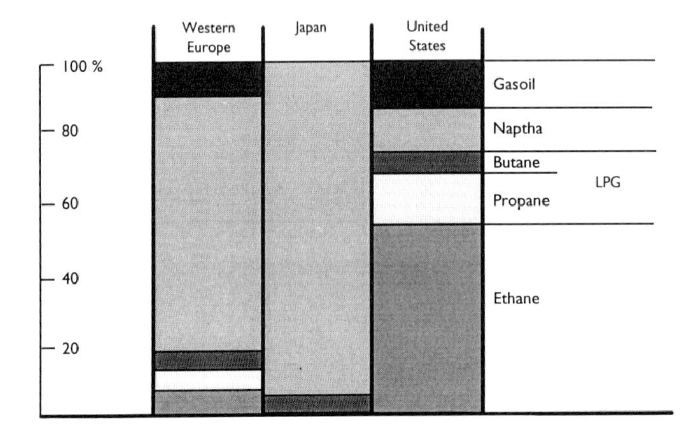

Source: Shell Chemical Information Handbook 1992

Manufacture of Base Chemicals

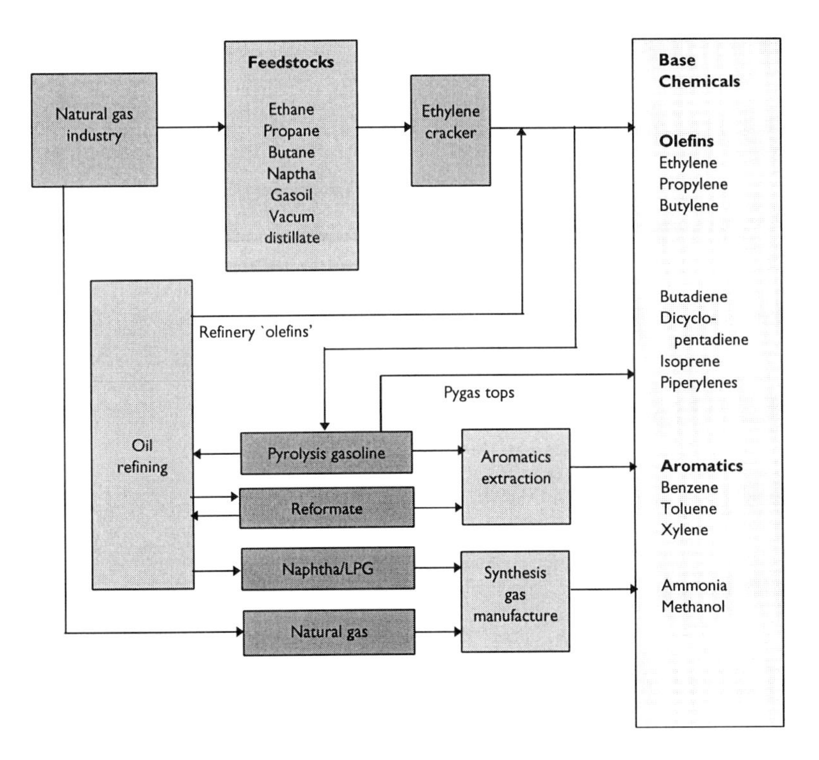

Source: Shell Chemical Information Handbook 1992

For Western European ethylene production the Association of Petrochemical Producers in Europe (APPE) has published a chart[16] showing actual raw material usage since 1981 with a forecast to the year 2000.

The major base chemicals are in two main groups.

One is called the 'olefins' or 'alkenes' comprising ethylene, propylene and butadiene. They are chemically described as 'unsaturated', containing a 'double bond' that forms a reactive centre of great value for synthesis of other materials.

Western European Ethylene Production by Raw Material

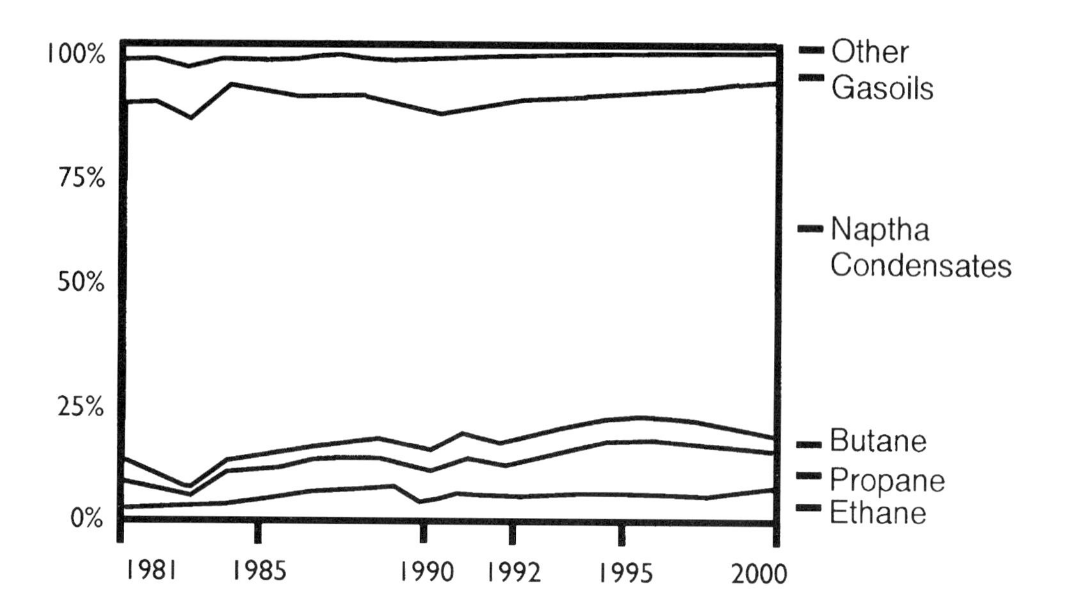

Source: APPE

The second group are the first members of the chemical series called the 'aromatics' comprising benzene, toluene and xylene, known as a group by their initial letters BTX. 'Aromatics' as a title should not mislead anyone into thinking that their smells are as fragrant as the first dictionary definition of the term suggests. The smells are unpleasant and the materials toxic. In modern chemistry, the term 'aromatic' means that the compound includes at least one six-membered benzene ring.

Other base chemicals are also of great importance. Methane, the main constituent of natural gas, is also used for synthesis. Acetylene and methanol are both long-established base chemicals; they were formerly obtained in other ways but are now mainly derived from petroleum and natural gas. In addition, the supply of ammonia which is of course not an organic compound, but is composed of nitrogen and hydrogen, is now based on natural gas and petroleum products which supply the hydrogen for the ammonia synthesis. In turn, the ammonia serves as a base chemical for fertilisers and for other chemicals.

Since nobody consumes these base chemicals as such, what is the reason for producing them, and how does the chemical industry come to be Britain's fourth biggest manufacturing industry? The clue lies in the conversion of the base chemicals generally through several intermediate stages into products of great utility to industry and to ourselves. These products are often so much better in their performance than direct natural products that they replace them; consequently the chemical industry tends to grow faster than the general economy as a whole[17]. Over the period 1982 to 1992 for example, the chemical industry in the UK grew at 1.9 times the rate of all manufacturing industry, in France at 4.1 times, in Italy at twice the rate, and in the European Union as a whole at 1.9 times the rate for manufacturing industry.

Examples of the recognizable consumer products derived from the base chemicals include, from the olefins, polyethylene, PVC, polystyrene, ethylene glycol coolant, synthetic rubbers and nylon. Often the final product has involved reactions with other base chemicals too, so that the BTX group also lead to polystyrene, nylon and synthetic rubbers, as well as producing latex, polyester fibres and other resins. Other derivatives of the base chemicals are applied in making pharmaceuticals, crop protection products, dyestuffs, adhesives, paints, cosmetics and a number of synthetic textiles. The base chemicals are as it were the hidden sources of a surprising range of familiar domestic materials.

An allied area of activity has been industrial 'farming' of micro-organisms on oil fractions to yield products of high protein content for feeding to animals. In 1959, a French BP team discovered that yeasts could grow well on waxy materials — chemically known as n-paraffins. During the following years processes were developed on to the industrial scale based on these waxes and separately on gasoil. Naturally products intended as foodstuffs had to be rigorously tested for nutritional value and for toxicity.

These remained minor activities and recent company reports[9] indicate that BP is organising what they call a 'managed exit' from this area, selling off these subsidiary businesses.

UK Trends

The discovery of North Sea oil and then the rapid development of production in the late 1970s and in the 1980s transformed the situation in the UK[5]. North Sea oil made major contributions to Britain's balance of payments and to government revenues. In the official statistics, figures are defined in terms of 'arrivals', indigenous production and shipments. Until 1973 arrivals had been increasing for many years but they peaked in 1973 then fell as production rose from 1974-75. Following the appalling disaster in 1988 on the North Sea oil rig called Piper Alpha, where an explosion and subsequent fire killed 170 people, there was a large fall in production (1988-89) due to both losing production facilities and to safety work. Oil production re-started to increase only in 1992. Understandably the trends in arrivals tend to be mirror images of those for home production, arrivals rising as production falls and vice-versa.

Turning to the uses of oil products, we find that in recent decades demand for fuel oil has sharply decreased and so has that for gasoil (generally products used as a heating fuel) other than Derv (diesel engine road vehicle) fuel. However, deliveries of motor spirit and aviation turbine fuel have steadily grown, and so have those of Derv but rather erratically.

Monitoring of emissions causing pollution of the atmosphere shows road transport as the largest single source (42%) of black smoke[5] — defined as fine suspended particles from incomplete combustion of fuel — with the domestic sector as the second largest at 35%. All other sources are a long way behind these two. Road transport is an even larger contributor (89%) to the growing emissions of carbon monoxide, also due to incomplete combustion of fuel, nearly all of this coming from motor spirit and hardly any from Derv.

In this area of damaging emissions, the most encouraging note is that emissions of lead into the atmosphere have fallen by about 80% since 1975, though petrol consumption has risen by about half. By April 1993 about half of all petrol used was unleaded.

According to a press review[8] the low oil prices causing difficulties elsewhere are being discussed by analysts in their implications for North Sea production. Several oilfields are considered to be under threat due to low margins between these prices and their production costs. Different analysts made contrasting predictions and there is no clear indication of probable trends but prospecting and development of oilfields continues implying continuing commercial confidence in the longer term future.

References

1 *Energy for tomorrow's world* London: Kogan Page for World Energy Council, 1993

2 *Our industry petroleum* London: British Petroleum Co plc., 1977

3 *BP Statistical Review of World Energy* London: British Petroleum Co plc., June 1993

4 On chemicals from petroleum there are several major textbooks and reference books, but a useful summary for the general reader is *Chemicals information handbook 1992* London: Shell Chemicals, 1992

5 *Digest of UK Energy Statistics 1993* London: HMSO, 1993

6 'Western oil and gas activity in the FSU has already taken off' *Petroleum Review* (London) 47(563) December 1993 pp.549-551

7 *Financial Times Oil and Gas International Year Book 1994* Harlow, Essex: Longman Group UK Ltd., 1993

8 Ruth Kelly 'Price slump spoils British celebration of its coming of age as an oil nation' *Guardian* (6 Apr 1994)

9 *Group Results January-December 1993* London: The British Petroleum Co plc., 1993; *Group Results January-March 1994* London: The British Petroleum Co plc., 1994

10 'Saudi plans to push light crude' *Technical Review Middle East* (March/April 1994) pp.43,44.

11 Roger Rainbow and Henry Tan *Meeting the demand for mobility* (Presentation to the Institute of Petroleum, London) London: Shell, December 1993 Shell Selected Paper

12 *Energy in profile* London: Shell, 1993 Shell Briefing Service no. 3, 1993

13 'Technology in the third millennium: energy' discussion meeting of the Royal Society and Royal Academy of Engineering, London, March 1994. For example paper by J Masters 'Oil and gas utilization: limits of efficiency and their impact on demand' published in D Rooke, I Fells, J Horlock *Energy for the future* London: E

and FN Spon for the Royal Society, 1995 Technology In The Third Millennium no. 6 ISBN 0419200509

14 BS 2000 *Methods of test for petroleum and its products* London. British Standards Institution, published in parts at different dates

15 BS 6843 For example, *Classification of petroleum fuels* London: British Standards Institution, published in four parts at different dates (Part 1: 1987 is equivalent to ISO 8216/1)

16 *The International Convention for the Prevention of Pollution from Ships; the 1978 MARPOL Protocol* and successive amendments. Summarised in 'MARPOL 73/78. The International Convention for the Prevention of Pollution from Ships, 1973, as modified by the Protocol of 1978 relating thereto' *Focus on IMO* (Jan 1994)

17 Association of Petrochemicals Producers in Europe, Av E Van Nieuwenhuyse 4, bte 1, B-1160 Bruxelles, Belgium

17 *UK chemical industry facts* London: Chemical Industries Association, June 1993

18 R R Chianelli 'Bioremediation: helping nature's microbial scavengers' in: *Proceedings of the Royal Institution of Great Britain* London: OUP/The Royal Institution, 1994 vol. 65 pp.105-126

19 *Gasoline* London: Shell Science & Technology/Shell International Petroleum Co Ltd., 1989

20 *Guardian* (16 May 1994)

Chapter 4

Gas

Contents

- Origins and Reserves
- World Trends; Production, Consumption and Trade
- Environmental Issues
- UK Trends
- References

Gas

"Theory of the true civilisation. It is not to be found in gas...."

Charles Baudelaire. *Mon coeur mis a nu.* LIX.

Seepages of combustible gas many thousands of years ago, probably first ignited by lightning, are thought to have been 'the fuel for the 'eternal fires' of the fire-worshipping religion of the ancient Persians'[8]. Many now again worship natural gas, though profanely as a clean premium fuel generating less CO_2 per unit of energy than other fossil fuels. More generally, in physics, the term 'gas' refers to a phase of matter where a substance can expand to fill any containing vessel. In discussing energy supply, the term refers to a fuel in this phase of matter — most often tapped from natural sources, and consisting mainly of the hydrocarbon methane or its close chemical relatives. Its principal use is as a source of heat or power but it is also processed to make carbon black, a large proportion of ammonia fertiliser, gasoline and liquefied petroleum gas.

Origins and Reserves

Gas is found in reservoirs alone (known as nonassociated gas) or associated with petroleum (associated gas) as briefly discussed in the previous chapter. It is considered to be derived from ancient land plants as well as organic residues from life forms that grew in water and is discovered below oil as well as within it and above it. Natural gas includes hydrocarbons in the gaseous state produced either from gas wells or produced with oil and separated from it. There may be higher members of the saturated aliphatic series present — the hydrocarbons ethane, propane, butane, pentane or even further higher compounds. Some natural gases, notably the French deposits at Lacq, have high sulphur content but more usually natural gases are free of sulphur. Many of the source rocks for significant gas deposits are said to be[8] associated worldwide with Upper Palaeozoic coal, in the region of 300 million years old.[8] Methane is also found trapped within coal seams and in many parts of the world is 'drained off' from the coal and used.

Other gases found with the principal hydrocarbons may include nitrogen, carbon dioxide, hydrogen, and helium or argon. Helium arises from radioactive decay of thorium and uranium, while much of the argon derives from the radioactive decay of a radioactive isotope of potassium; both of these gases are thought to be coincidentally caught in the same trap formation as the hydrocarbon gases.

When the principal hydrocarbons occur in liquid form — either by human action or in nature — they are referred to, slightly confusingly, as 'gas liquids'. These comprise; —

LNG — liquefied natural gas consisting mainly of methane, CH_4, chemically the lowest member of the saturated aliphatic series;

NGL - natural gas liquids, consisting of ethane, C_2H_6, chemically the next higher member of that series, mixed with heavier fractions, all of which have much higher boiling points than the methane and are extracted from the gas;

LPG - liquefied petroleum gas, consisting of the next higher members of the saturated aliphatic series of hydrocarbons, propane C_3H_8 and butane C_4H_{10}.

From time to time there are also inconclusive speculations about methane generated by inorganic processes connected with the earth's formation, since hydrocarbons are found in carbon-containing meteorites of what is called the chondrite type. These are stony meteorites containing small mineral granules.

About 40% of the world's gas reserves are in the former Soviet Union where, despite the political turmoil of recent years, amounts of proved reserves continue to increase. Iran accounts for 14% of the world total[1]. Though gas reserves are considered to be abundant, an appreciable proportion of them are concentrated in a few countries. About a third of the total are in 11 giant fields[2]; six are in the former USSR, the others in Qatar, Iran, Algeria, the Netherlands and Norway.

Continued prospecting, has resulted in continuing increases in proved reserves at rates that are far faster than the increases in production and consumption. So the ratios of proved reserves to current production, known as the R/P ratios and measuring the years of reserves remaining, have risen from below 40 in 1966 to almost 70 in 1992 — much higher than those for oil and growing much faster. The number of countries known to have worthwhile reserves of gas has more than doubled in the last thirty years to 85 today. Geologists comment that there are substantial volumes of untested sedimentary areas worldwide. These are areas that were formed by matter that was carried by water or wind and deposited and may then have been consolidated into rock and may hold gas. So there are thought to be prospects for discoveries of many further major gasfields[6]. Natural gas has been found in every system of rocks down to the Cambrian, that is to the Period between about 590 and 505 million years ago.

However, despite these apparently dazzling prospects for future supplies, researchers believe that there will be pressures on supplies during the next century[7], and research continues on gasifying coal which has enormously greater reserves.

Proven Natural Gas Reserves

Total: 131 500 milliard m^3 @9500 kcal/m^3

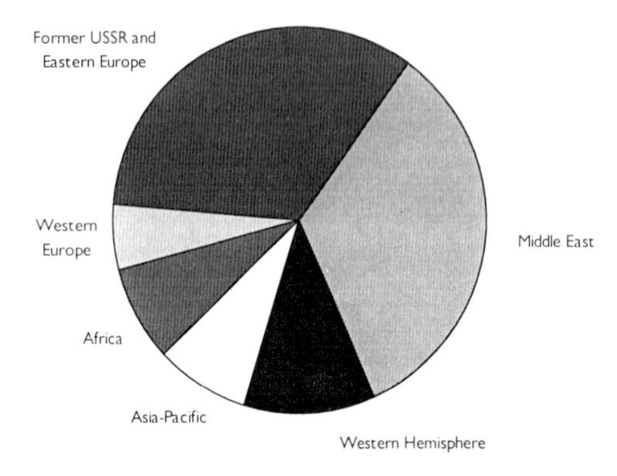

Source: Energy in Profile. Shell

World Trends; Production, Consumption and Trade

Though production in 1992 was only about 0.4% greater than that in the previous year, the trend for several decades has been one of strong growth but in the former USSR, though it is still the world's largest producer, there was a decline of 3.7% largely due to a fall in output from Turkmenistan. In the USA, the second largest producer, there was also a small fall in production. Despite these falls, these two powers together continue to yield over 60% of world gas. High percentage increases were shown by Canada, the Middle East and by several countries in Asia, the highest being that of Malaysia though it is a relatively late starter.

Early uses of fuel gas did include natural gas but gas formed by heating coal in the absence of air (destructive distillation or carbonisation) became a primary fuel in Europe for lighting from about 1790. With many improvements in technique in production, distribution and use, this coal gas continued in use for lighting and heating for more than 150 years. Where it is now used, it tends to be a by-product of some other activity such as producing metallurgical coke. Coal gas has now generally been replaced by natural gas, which has a calorific value of over 38 MegaJoules per cubic metre (MJ/m^3), more than double that of gas made from coal; unlike the latter, natural gas is free of the poisonous carbon monoxide which too often caused fatalities.

Natural gas is generally considered a convenient and clean fuel. Since the shortages of oil that developed around 1970 gas has become an important source of energy on a world scale and though world consumption rose only a little in 1992 compared with 1991, this average effect resulted from a variety of changes in various areas. In the former Soviet Union usage fell, and of its former associated countries only Poland showed a marked increase. US and Canadian consumption rose, and there were relatively large increases in India, Malaysia, and South Korea, though these were small as shares of world usage. A striking feature is the growing use of gas for generating electricity, a trend expected to continue. This is seen as the fastest growing market for gas in Europe with significant increases[3] in UK and Italy, but also elsewhere as in Japan and other parts of south east Asia. Gas industry spokesmen naturally emphasise the environmental advantages of this fuel and its suitability for burning in technologically advanced combined cycle generating plants[3]. In these, the discharge heat from one power generator, usually a gas turbine, is used to generate further power, by raising steam to drive a steam turbine. However, other fuels can also be used in combined cycle generating plants and can be made to meet environmental requirements as we have noted in earlier chapters; so the choice of energy supply in a particular case needs a specific study which takes into account relevant local conditions.

Challenging the worldwide dominance of liquid petroleum fuels, natural gas is also being used to power vehicles in many countries. The largest users are reported to be the former Soviet Union and Italy and it is estimated that there are now well over one million of these vehicles on the road. Interest seems to be increasing particularly on grounds of reduced atmospheric pollution compared with the emissions noted earlier from vehicles powered by petroleum source liquids. Progress in this application depends on policies of governments, since the results of any attempts at economic comparisons turn heavily on the expected tax regimes for the respective fuels.

There has long been interest in large-scale gas storage to balance the loads on production between seasons, since demand is lower in summer and higher in winter. For the UK, British Gas state[9] that winter demand can be six times that in summer. Storage possibilities include —

1 as LNG at about minus 160 C which reduces the volume to about one six hundredth of that of the same amount of dry gas, in specially designed LNG gas storage holders;

2 in underground impervious caverns such as worked-out former salt deposits;

3 in exhausted gasfields;

4 locally in the familiar low pressure gasholders often confusingly called gasometers;

5 in high pressure stores that include the distribution pipelines themselves, since gas is distributed over longer distances at high pressure. British Gas states that it uses compressor stations to build up the pressure to 75 bar — roughly 75 atmospheres pressure.

Storage is now reported[2] as an increasingly important factor in the USA where 'capacity is believed to be almost half of total annual production.'

Though countries do use varying proportions of their own gas deposits, a large part of the world gas reserves are a long way, often thousands of miles, from the large-scale consumers. But the 'thermal density', the energy value per unit volume, of gas is only about 1/900 that of oil. It is for this reason that gas, associated with oil, has often been flared off, burnt, apparently wastefully. About 10% of world annual gas production was estimated as being lost at the wellhead in 1980 by this procedure[8], Middle Eastern and African oil producers having flared the most gas. Some was used by reinjection to build up pressure in the wells and the rest flared. In the UK one of the conditions of the petroleum production licences is that gas may be flared only with the consent of the Secretary of State — obviously intended as a pressure on operators to promote conservation.

In the past where gas was used, particularly in the nineteenth century the consumption was localized because there was then no way to transport it over appreciable distances. So gas has to be concentrated, by compression or by liquefaction, to make its transport economic. Turning the gas into liquid of course greatly reduces its volume as we have noted and therefore increases the thermal density. The table below[5] indicates that methane, the main constituent of most natural gas, cannot be liquefied without substantial cooling even if high pressures are applied.

Significant temperatures

	Boiling Points at Atmospheric Pressure °C	Temperature above which gas cannot be liquefied °C
Methane	- 162	- 82
Ethane	- 89	32
Propane	- 42	97
iso-Butane	- 12	134
n—Butane	- 1	152
Pentane	36	197

Thus only pentane (C_5H_{12}) and higher homologues can be transported, with the necessary safety precautions, in fairly standard vessels. (The term 'higher homologues' means compounds that are like the original except that they have larger numbers of carbon atoms such as hexane (C_6H_{14}) and heptane (C_7H_{16}) and are of higher boiling point). Despite these difficulties, large and increasing volumes of natural gas are traded by pipeline and as LNG amounting in total to 16% of total world consumption in 1992. In that year the largest exports were by pipeline from the Soviet Union to several countries in Europe both members of the OECD and others. The former Soviet Union also holds the record for the longest gas pipeline. This is a 5470 kilometre long Northern Lights pipeline that crosses the Ural Mountains and some 700 rivers and streams, linking Eastern Europe with the West Siberian gas fields on the Arctic Circle[8]. Thus gas from the Urengoy field, the largest in the world, passes to Eastern Europe and on to Western Europe. Netherlands and Norway are also large suppliers to OECD European countries, but the UK has very small exports and even takes a portion of its supplies from Norway. Massive Canadian exports by pipeline to the USA made it the world's second largest exporter.

Overall about three quarters of world exports are by pipeline and the balance in the form of LNG. Trade in LNG is dominated by Indonesia which supplies about 40% of these exports, most of it going to Japan. Other large LNG exporters to Japan are

Malaysia, Brunei, Australia and Abu Dhabi, with a surprising though relatively small contribution from the USA. Algeria supplies large quantities as LNG to several West European countries, France taking the largest quantities, and also by pipeline via Sicily to Italy. Construction of this pipeline was described[8] as being of great engineering difficulty. The sea is more than 610 metres deep along some parts of the route. Gaz de France has announced that it is considering co-operating in a major pipeline from Algeria, through Morocco, across the Straits of Gibraltar and across Spain.

Estimates that the global consumption of gas will exceed that of oil on a thermal basis in the first decade of the next century[5] has led to increasing interest in the

Natural Gas Trade, 1992

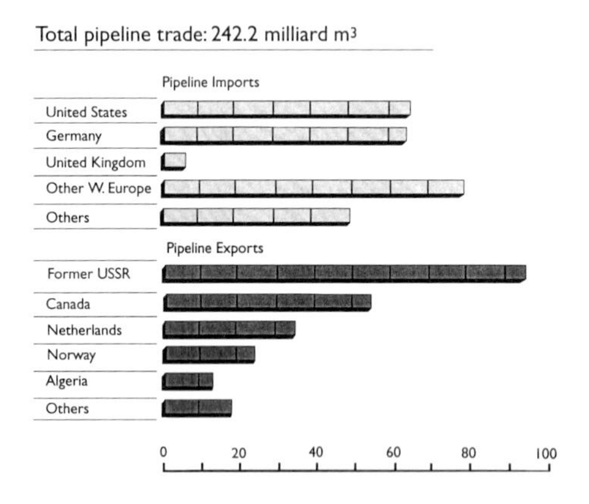

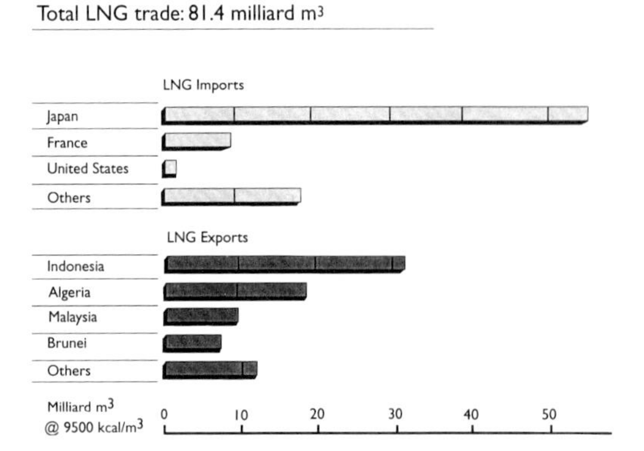

Source: Energy in Profile No. 3, 1993. Shell

prospects of supplying gas to Europe from the Middle East, where exports have so far been minimal. Current forecasts are that there may in due course be supplies from Iran by pipeline, and from Yemen, Oman and Qatar in the form of LNG.

Heavy investment is going to be needed whichever system of delivery is used, to find, produce and transport the greater quantities of 'new gas' to their user countries. There has naturally been steady improvement in the design and reliability of liquefaction plants and in their speed of construction so that unit costs of liquefaction have been reduced. But after allowing for these improvements the Shell Managing Director responsible for gas put the figure[3] for the aggregate investment needed at up to $110 billion over the next twenty years, plus of course the needs of the local distribution systems in the receiving areas.

Since natural gas is potentially an explosive material, it is highly relevant to add that tanker designers and operators have worked to very strict standards of safety; the industry claims that there have been 'no major incidents compromising safety in over 15,000 voyages' to the beginning of 1993. However, there were some accidents with static storage tanks in the earlier days. Preventing leakages or spills of LNG is of great importance since the danger caused may not be transient. LNG is liable to form a cloud of cold gas that hovers near the earth and does not readily disperse in relatively still air but most static installations now use double wall metal tanks with powder insulation between the walls. Of more than 200 of these storage facilities, it is stated that 'none has had a serious fire or other problem during operation'[6]. Suitably designed road tankers may be used to move LNG and supply it to points not yet connected to the distribution system or temporarily isolated for maintenance or repair.

Environmental Issues

Fuel gas is obviously a much cleaner material to deal with than coal or fuel oil, despite the big advances in preventing pollution from the latter fuels. Gas industry interests understandably boast that its distribution by pipeline causes no visual intrusion; it avoids adding burdens to conventional transport systems (with their environmental impact) except where the use of LNG transport is unavoidable.

Amid current anxieties about global warming and the need to reduce the generation of CO_2 the advantage of natural gas is particularly striking, justifying its claim to be the 'greenest' of the fossil fuels due to its higher proportion of hydrogen and lower proportion of carbon. Figures for generation of CO_2 calculated on a comparable basis have been kindly supplied in 1994 for the UK by Mr George Henderson of the Building Research Establishment. The figure for CO_2 represents the kilograms (Kg) emitted per Gigajoule (Gj) of energy source delivered at the user's door or meter.

FUEL	Kg CO_2 emitted per Gj of energy
Gas	52
Heating oil	75
Liquid petroleum gas	76
Household coal	83
Anthracite	90
Smokeless solid fuel	116
Electricity	188

The figure for electricity was averaged over all sources of electricity and showed a big reduction over earlier estimates in 1987. Figures for other countries are likely to differ only in detail.

UK Trends

The UK was one of the pioneers in manufacturing and distributing coal gas first for lighting and later for both heating and lighting. The gas in due course became known as town gas since other fuel gases — such as that from gasoil mentioned earlier — were mixed in when necessary to maintain the statutory level of calorific value, monitored by Gas Inspectors. Although the gas is measured (metered) by volume, charges are based on the heating value delivered measured by the calorific value per unit volume under standard conditions of temperature and pressure.

In 1949 the industry in the UK, consisting of a considerable number of gasworks and localized networks, was nationalized under the Gas Act 1948. After further administrative changes, it was then privatised in 1986, becoming British Gas plc. This company remains the largest gas supplier in Britain; it also produces gas offshore and onshore, conducts exploration for gas and has a range of related activities worldwide. To transmit and distribute the gas in Britain it operates a pipeline system with a total length of 240,000 km, but it is obliged to act as a common carrier for other gas supply companies on commercial terms. As a public gas supplier British Gas, under the Gas Act 1986, is subject to monitoring by the Office of Gas Supply (OFGAS) to ensure that it meets the conditions of the company's authorization. OFGAS also grants the authorization[13] to other suppliers of gas through the pipes, fixes and publishes maximum charges for reselling gas and keeps under review conditions that may affect competition. By the middle of 1994, some 60 companies were listed by OFGAS as authorised to supply gas under the common carriage arangement with British Gas. Users of more than 2,500 therms of gas (73,200 kWh) per annum are eligible to apply to one of those companies for a gas supply. Separate application is needed for supplying gas through pipelines belonging to the shipper himself, or to a third party. Authorisation is not needed where a company is supplying only itself or another company in the same group. There are also special arrangements in Section 6 of the Act for supplies of above two million therms a year; in these cases a simple notification to the Secretary of State for Trade and Industry is sufficient. For other companies, OFGAS requires evidence of the financial status of the company, of safety arrangements, together with details of the premises supplied and the expected annual rate of usage. In effect, the Gas Act 1986 introduced competition in the industrial and commercial sectors but new developments, heralded in the Queen's Speech, in autumn 1994, will extend competition to the sector suplying domestic premises. From 1 April 1996 it is intended[16] to operate a two year pilot scheme along these lines in selected areas. If the trial is successful, the scheme will then be extended nationally.

Supply of town gas rose slightly up to the end of the 1960's then fell, contributing to the decline of the UK coal industry, as town gas was replaced by natural gas in a major nationwide operation that involved changing jets on existing equipment so that all gas-fired apparatus would burn the new fuel satisfactorily. In some cases complete combustion units had to be changed. Measured by the energy value, the use of natural gas[10] has increased in recent years to a level more than five times that of town gas in the peak in the late 1960's.

Some 90% of the total natural gas available is from indigenous production — from the North Sea and onshore sources. The balance of natural gas is imported from the Norwegian sector of the North Sea though the proportion of imports has been falling. In the mid-1980s imports were about 33% of indigenous production. By 1992 this proportion had fallen to roughly 10%.

About 15% of the total gas consumption in the UK is from a variety of other sources. Town gas totally disappeared from the statistics after 1988. There are appreciable contributions from coke oven gas, where the fuel gas is a by-product of coke manufacture and from blast furnace gas, also a by-product, in this case from the smelting of iron ore. Much smaller amounts come from colliery methane piped up to the surface from coal seam 'drainage', and from landfill gas and sewage gas both of which will also be considered later among the 'renewable sources'.

Of these lesser contributors to the UK gas fuel supply, the two largest are listed as 'liquefied petroleum gas' and 'other petroleum gas'. The liquefied petroleum gas production is propane and butane coming in part from petroleum refineries, as earlier mentioned, with the balance from the same compounds originating in North Sea associated hydrocarbon gases obtained at gas separation plants. The other petroleum gas production comprises tail gases from refineries, other than propane and butane, plus ethane from the North Sea associated hydrocarbon gases, also from the gas separation plants.

About half of the total gas available was used in the home and the remainder shared between industrial (including petrochemical feedstocks) and 'other' uses. The last category includes fuel producers, public administration, agriculture and commerce. The 'dash for gas' in generating electricity led to a sharp increase in the use of gas for this purpose but by 1992 this was still only a small absolute quantity and a small proportion — equivalent to 1 million tonnes of oil (Mtoe) out of a total of 70.8 (Mtoe) total UK fuel input for this purpose.

Further large increases were however planned for the future. By 1993 the share of gas in generating electricity had risen[15] to 10% of the fuel input and National Power had announced a programme to upgrade its generating capacity with some 5000 megawatts of combined cycle gas turbine plant by the end of the decade.

British Gas maintains a large active R & D programme, often in collaboration with manufacturers. They cover such themes as improvements in burners, furnace design, instrumentation and computerized control systems. Other areas are improvements in incinerating waste, in gas engines and in converting standard car engines to operate on gas, which also involves work on the best way of storing that energy source. As well as programmes on combined heat and power based on gas turbines, there is also one on fuel cell power plants. These cells are devices that convert fuel directly into electricity, an apparently simple concept that has proved to be difficult to translate into economic and practical plant. Like the coal researchers, those working with gas see good prospects for fuel cells — the coal researchers working on gas from coal by modern processes[11], while the gas researchers apply natural gas directly. All major gas projects are stated[12] to be subject to environmental assessment at a very early stage and guided by an Environmental Planning officer.

References

1 *BP Statistical Review of World Energy* London: British Petroleum Co plc., 1993

2 *Energy in profile. Number three, 1993* London: Shell Briefing Service. Shell International Petroleum Co Ltd., 1993

3 J R Williams *Natural gas - the world supply challenge* London: Shell International Petroleum Company, 1993 (An address to GASTECH 93 Paris, February 1993)

4 Peter Jones 'Global demand for gas grows' in: *Technical Review Middle East* (January/February 1994) pp.33-36

5 *Our industry petroleum* London: British Petroleum Co Ltd., 1977

6 'Liquefied natural gas (LNG)' in: *McGraw-Hill Encyclopaedia of Science and Technology* 7th ed., New York: McGraw-Hill, 1992 vol. 11

7 *Energy for tomorrow's world* London: Kogan Page for the World Energy Council, 1993

8 'Natural gas' in: *New Encyclopaedia Britannica Macropaedia* London: Encyclopaedia Britannica Ltd., 1993 vol. 19

9 *British Gas educational resources* London: British Gas Education Service, 1994 (This is a catalogue listing a wide range of publications on many aspects of the industry)

10 *Digest of United Kingdom Energy Statistics 1993* London: HMSO, 1993

11 David H Scott *Advanced power generation from fuel cells — implications for coal* London: IEA Coal Research, 1993

12 PF Weatherbilt *Major British gas projects and the environment* Pembroke College, Oxford: Institution of Gas Engineers, 1990

13 *Authorisation under Section 8 of The Gas Act 1986. Notes for applicants supplying under common carriage arrangements with British Gas plc through the British Gas pipeline system* and allied documents London: Ofgas

14 George Henderson *Private communication* Building Research Establishment

15 *Digest of United Kingdom Energy Statistics 1994* London: HMSO, 1994 p150

16 Private Communication from Office of Gas Supply, 105 Victoria St, London SW1E 6QT

Chapter 5

Nuclear

- Nuclear Reactions
  - Fast Reactors
- Nuclear Reactors: Capital Requirements
- Reserves — and Waste Disposal
- Decommissioning of Nuclear Reactors
- Environmental and Safety Issues
- World Trends
- Fusion
  - Cold Fusion
- UK Trends
  - Radiation Exposure of Public
  - Occupational Exposure of Radiation Workers
- References

Nuclear

"There is no evil in the atom; only in men's souls"

Adlai Stevenson. Speech 'The Atomic Future' 18 Sep 1952 in *Speeches of Adlai Stevenson* 1952.

Attitudes towards civil nuclear power form one of the great schisms of mankind. Many people, probably still influenced by memories of its early gestation from work on the devastating nuclear bombs, regard this source of power as little short of the work of the devil. These apprehensions based on older events were reinforced by the disasters at Three Mile Island (USA, 1979) and the more serious one in Chernobyl (Ukraine, 1986) which are often re-quoted. Others see nuclear energy as a major contributor to the growing need for energy supplies with the added benefit of generating neither CO_2 nor gases producing acid rain. Future risks are seen as calculable and very low — an important topic further discussed below.

Nuclear Reactions

Traditional fuels — wood, dried dung or other biomass, peat, coal, oil, gas — deliver their energy by combining with oxygen in a chemical reaction. Atoms, the smallest unit of an element, are now widely understood to have a positively charged nucleus surrounded by negatively charged electrons. Atoms of different elements differ in their mass, the size of positive charge on the nucleus, and in the number of electrons. In chemical reactions, atoms combine with each other. The atomic nuclei remain unchanged but some of the outermost electrons are removed, transferred or shared.

Nuclear reactions involve the nucleus of the atom. These reactions are usually accompanied by energy changes that are millions of times larger weight for weight than those of ordinary chemical reactions such as the burning of fuels. (In practice, the amount of energy obtained weight for weight is tens of thousand times as much as that from chemical reactions.) The nuclear reaction may be spontaneous. Radioactive elements — the classic example is radium — undergo spontaneous nuclear transformation needing only a single nucleus; (the radium 'decays' forming another element, the gas radon which is itself radioactive.) Other nuclear reactions may occur when nuclei are bombarded with particles such as neutrons, protons and others. (Neutrons are elementary particles without an electric charge, while protons are elementary particles of roughly the same mass as the neutron but with a positive electric charge equal in magnitude to the negative charge of an electron.)

Nuclear energy may be liberated by a fission nuclear reaction where an atom splits into lighter atoms; or by a fusion nuclear reaction where atoms truly fuse together; or by radioactive decay. The energy liberated by these reactions is equivalent to a slight loss of mass in the system. The quantity of energy is given by the apparently simple relationship $e = mc^2$; in this the letter e represents the energy, m the mass transformed and c the speed of light.

Current nuclear reactors are based on fission of one form of uranium, known as U235, or plutonium in a controlled nuclear chain reaction, but there are major research projects on nuclear fusion. In the controlled nuclear chain reaction a neutron hits a fissile atom that splits releasing energy and more neutrons. The neutrons collide with other fissile atoms, continuing the process, maintaining the 'chain', that is the self-sustaining nuclear reaction. So a reactor needs[9] a neutron source for starting it up, a collection of fissile material as fuel, a neutron absorbing material for controlling the rate of reaction and a coolant that takes heat from the reactor. It is the heat available in this coolant that is then used to generate steam that drives the turbine, and in turn the electric generator.

Naturally not all the neutrons produced in the atomic fissions succeed in producing further fissions. Some are absorbed in the reactor structure, or escape from the reactor and are absorbed in the shielding. There is a 'critical mass' or 'critical volume' which is the minimum mass or volume of fissile material, that is capable of a self-sustaining chain reaction.

Then, even if the neutrons are absorbed in further fuel, they do not all result in further fissions. The neutrons resulting from fission are known as 'fast neutrons' — travelling with high speed, typically about 20,000 km per second. Natural uranium consisting of 0.7% of U235 and 99.3% U238 cannot sustain a chain reaction with these fast neutrons. There is only a low probability of their being captured and resulting in fission. So the fast neutrons are slowed down by repeated collision with the atoms of a 'moderator' such as graphite, water or heavy water, to a speed of about 2.2 km per second. They are then described as 'thermal neutrons' and the probability of causing further fission with U235 becomes several hundred times greater than that of any neutron, fast or slow, with U238. There are also technical advantages in increasing the proportion of U235 in the uranium. This is called 'enrichment'. The reactors with moderators are known as 'thermal reactors' and there are several variants.

Yet some enriched fuels — notably mixed plutonium-uranium fuel produced in the first place as a by-product from thermal reactors — are able to sustain the fission reaction caused by fast neutrons. Reactors of this type are called fast reactors but they are still subjects of major and very expensive research projects. They need no moderator, have a more compact core than thermal reactors, and use liquid sodium metal to transfer the heat. The core can be surrounded by uranium (called a 'uranium blanket') which is partly transformed into plutonium. So this type of reactor 'breeds' further fuel and is also called a 'breeder reactor'. It produces about 60 times as much energy from the fuel as the best fission reactors.

Most of the world's power reactors are thermal types. In these the fuel is most often enriched uranium where the proportion of U235 has been increased by suitable processing to around 3%. The most widely used civil nuclear power reactor in the world is the Pressurised Water Reactor (PWR) design. In this, light water at high pressure is heated within the reactor core (the central region with the fuel elements where the chain reaction takes place). The heated water passes to heat exchangers where it boils water circulating in a separate circuit that is used to drive the turbines.

'Light water', mentioned above, is in fact ordinary water but in nuclear technology it is so-called to distinguish it from 'heavy water' where the hydrogen of the water is heavy hydrogen, also known as deuterium. Some reactor designs use heavy water as a moderator. There are also many hundreds of reactors powering naval vessels including submarines.

Fast Reactors

In France a fast reactor was due to be restarted in 1994 on the condition that it "...will no longer be used as a nuclear power station but will become a reactor devoted to research and demonstration"[6]. This reactor, called Superphénix, is due to develop methods of operation at low power to limit the quantity of plutonium produced. It will also be used to explore the means to destroy plutonium and high activity waste. These requirements were set following advice given by the French nuclear safety authority (DSIN) and the report of a commission set up to carry out a public enquiry. Within this framework, the owners plan to demonstrate the capacity of Superphénix to produce electricity at an industrial level.

Britain has been involved in fast reactor development since the early 1950's.[11] Design studies were followed by a Zero Energy Fast Reactor, then construction of the Dounreay Fast Reactor rated at 60 MWt, 15 MWe. (MWt is MegaWatt thermal, and MWe is MegaWatt electric power which is only a fraction of the heat generated). Construction of the Prototype Fast Reactor (PFR) was the next stage, also at Dounreay. This larger reactor, designed to produce 250 MWe from 600 MWt core power achieved full power in 1977. Parallel developments had been in progress in Russia, the USA, France, Germany, Japan and Italy with contact between researchers that has resulted in joint R & D programmes. Britain is also involved in European joint activities including the European Fast Reactor Utilities Group, and the European Fast Reactor Associates (EFRA) design consortium. All this coordination has resulted in design work of EFRA being directed towards producing a single, agreed design for a European Fast Reactor. This is intended to incorporate the best features of three former 'national' designs; they are the UK Commercial Demonstration Fast Reactor (1250 MWe), the French Superphénix 2, and the German SNR-2. The wider contacts have also been maintained.

The UK government decided to end PFR operations early in 1994, but Japan is continuing with its programme and Superphénix is restarting operations as noted above. The UK Atomic Energy Authority (AEA) has re-stated its view[11] that the fast reactor is the system of the future. Much more energy can be produced by this means from uranium than can be achieved with the best thermal reactors, by converting U238 that is otherwise unusable into plutonium.

Nuclear Reactors: Capital Requirements

Contesting the view often mooted that nuclear power has exceptional capital needs, Dr R Hawley of Nuclear Electric has put forward this comparison[10]. In assessing the capital for a combined cycle gas turbine plant (CCGT) the cost of developing the gas field must also be considered and added in. Similarly for a coal-fired station the capital cost must also allow for that of the deep mines needed. Renewables too have high capital costs. So, considering the capital needed to match by other means the output of a modern 2,800 MW twin PWR which is about 19 TWh per year, results in the following table. (TWh is 1012 Watt-hour, or million MWh).

Capital in Billion Pounds Needed to Generate 19 TWh/year

	Plant	Fuel Resource	Total
Gas	1.4	2	3.4
Coal	2.2	1	3.2
Nuclear	3.5	up to 0.3*	up to 3.8
Wind	7.5	-	7.5
Tidal	10.6	-	10.6
Solar	47	-	47

*This assumes reprocessing and MOX (Mixed oxide) recycle

On the basis of this argument, total capital is comparable with that for gas or coal — though oil has been omitted from the discussion and the pattern of investment needs over time is very different for the fossil fuels on the one hand and the nuclear on the other. Furthermore, the costs of decommissioning for the different kinds of plant have been omitted, but considering building costs on the basis stated, nuclear is only a fraction of the cost of the renewables on an equal energy basis.

Reserves — and Waste Disposal

Discussion of reserves is always beset by questions of definition. In the case of the fossil fuels, we have noted that proved reserves are taken to be those quantities of the fuel that are economically recoverable; more formally this is defined as the quantities that geological and engineering information indicate can be recovered from known reservoirs or deposits under existing economic and operating conditions.

This naturally applies to the uranium needed for fission reactors but there are also other issues.

Uranium and plutonium from spent reactor fuel can be recycled and used as Mixed Oxide (MOX) fuel. With fast breeder reactors, uranium reserves could last for over 1000 years[10]. On this basis, the AEA quotes[11] the current stockpile of U238 that is a by-product of the cycle of activities connected with thermal reactors, as the UK's largest indigenous energy reserve. Though the fission of the uranium isotope 235 produces a mix of 34 different elements, spent nuclear fuel overall consists of about 96% uranium, 1% plutonium and 3% waste material.

A further large reserve has been provided by the ending of the cold war and the reduction in numbers of nuclear weapons yielding plutonium and highly enriched uranium. Using these quantities of highly enriched uranium (HEU), after dilution, as fuels for PWR units would truly be turning most destructive swords into ploughshares. The Russian and US inventories of plutonium, used in Mixed Oxide Fuel[10], could supply three years of fuel for the world's PWR stations. A US study[24] warns of the danger of the surplus fissile material and urges two promising alternatives. They are to burn the plutonium in nuclear plants or to vitrify it into glass logs that would include high-level waste.

In the hope of defining the situation about uranium supply and demand the Uranium Institute has conducted a detailed worldwide survey of utilities and producers. Its report[26] acknowledges that the prospects for nuclear power growth and for the future uranium market remain uncertain. Yet it expects that after allowing for demand being met from current planned primary production, HEU, recycling and inventories, there will be a need for some limited new uranium capacity beyond what is now planned.

Despite public disquiet over the safety of waste disposal, the industry appears confident that it can handle the problems. It contrasts[19] the small quantities involved with the enormous quantities resulting from domestic waste, industrial waste, toxic chemical waste, coal mining waste, and fly ash from coal-fired power stations.

Radioactive waste is grouped under the categories low level, intermediate level and high-level waste. Though at present wastes are stored in most countries with due safeguards on the surface, underground storage has now generally been adopted worldwide as the desirable policy which is being introduced. Some low and all intermediate level waste will go into stores with several barriers between it and the outside world. A typical scheme involves waste put into concrete in steel drums, in an overpack, buried in an underground vault, which is then filled and sealed.

High level waste gives off heat and will therefore continue to be kept for perhaps fifty years in well-protected stores to decay and cool. The relatively small volumes are being converted into a glass. Radioactive waste disposal is an area of major international collaboration with the International Atomic Energy Agency, the EU and the OECD all involved. Most elaborate of the disposal sites is considered to be a deep repository in Sweden near the power stations at Forsmark on the Baltic. Research worldwide on this issue has covered geological stability, possibility of human intrusion, climate change such as a future Ice Age, microbiological activity, and behaviour of burrowing animals. In the UK, the Royal Society has carried out a review of the issues involved in disposing of radioactive wastes in deep repositories. It concluded[25] that this is a very demanding task needing further research but considered that individual items of the scientific work of the UK organisation (NIREX) charged with this responsibility, are of high quality and command worldwide respect. The issues are further discussed in a later section of this chapter dealing with the UK.

Decommissioning of Nuclear Reactors

A further area where there are big differences between the views of the industry and public perceptions, is in the decommissioning of nuclear reactors. Techniques have been developed so that several such reactors and radioactive plants have in fact been successfully dismantled and decontaminated. The industry emphasises that the recommendations for delay in dealing with the most highly radioactive parts of plants are not excuses for procrastination but practical measures to allow radioactive decay and make the job simpler.[10, 20] After the plant is closed down, nuclear fuel and some other radioactive material are removed. Surrounding buildings are dismantled and the reactor itself — occupying about the area of a football pitch — may be covered over.

By operations essentially of this kind, the world's first prototype PWR at Shippingport in the USA has been decommissioned and cleared, while the Lucens reactor in Switzerland is described as having been 'completely dismantled'. A 100 MWe prototype reactor in the nuclear research centre of Karlsruhe, Germany has been 'disassembled'. This has provided important demolition expertise for handling commercial power plants in the future.[27] In Britain the Windscale prototype AGR is being decommissioned; further similar work on three Magnox reactors is due to be treated as a group of demonstrations intended to build public confidence. France has taken this idea a stage further, turning the decommissioned Chinon reactor into a museum and activity centre attracting 25,000 visitors a year.

Environmental and Safety Issues

Nuclear energy can play a major part in meeting the growth in energy demand that we examined in the first chapter without adding to the atmospheric burden of CO_2 or acid gases. Contrary to views widely held among the general public, civil nuclear power contributes very little to ambient levels of radiation. In addition, workers in the industries tend to be kept under regular health surveillance. Favourable results of continuing surveys of radiation exposure for both the general public and radiation workers in the UK are discussed below.

In respect of the possibilities of nuclear catastrophe "the perplexing problem...is the quite extraordinary gap between the precisely calculated risk assessments of scientists and engineers and the perceptions of ordinary people" wrote Professor Terence Lee[22]. Apart from the frequent comparisons with large numbers of premature deaths due to smoking, and the accidental deaths due to motorcycling, it is also possible to note such areas as deaths in swimming and in horse-riding that are treated as 'acceptable' and do not result in calls to close down the activity. Yet this proposal is made in respect of the nuclear industry although there have been no deaths directly attributable to nuclear power in the UK. The industry reasonably claims[10] that it operates standards of safety unrivalled by any other sector of the energy industry. The Chernobyl accident and its resulting human tragedies, occurred in a reactor design rejected as unsafe in the UK and in other countries outside the former Soviet bloc. International technical assistance is now being provided to improve the safety of those reactors, as noted below. It is likely that international financial co-operation will follow. There is now a fully comprehensive World Association of Nuclear Operators promoting the highest standards of operational safety throughout the world, and risk assessment has become a growing area of scientific study[23].

Critical environmental and anti-nuclear groups such as Friends of the Earth, Greenpeace and the Campaign for Nuclear Disarmament (CND) maintain that the claims of safety and cleanliness are fraudulent[28]. Current generations, they say, have no right to impose on future generations the legacy of radioactive waste that remains hazardous for hundreds of thousands of years. Further objections are based on the radiation exposure of nuclear workers, discharges of radioactivity into the environment, risks of disastrous accidents and of nuclear weapons proliferation. They urge greater attention to energy efficiency and to using renewable energy sources to eliminate the need for nuclear power.

World Trends

The USSR put their first power reactor into operation at Obninsk near Moscow in 1954[1] and two years later a British reactor at Calder Hall began to generate electricity. Though there has always been some degree of confusion about facts and figures[16] since countries were also building and operating reactors to produce military grade plutonium, in due course several countries began to use civil nuclear reactors so that the amount of power generated in this way has steadily increased until 1991, then stagnated during the following year[2].

Following a UN International Conference on the Peaceful Uses of Atomic Energy, the International Atomic Energy Agency (IAEA) was set up in 1957 reporting annually to the UN[3]. It aims 'to accelerate and enlarge the contribution of atomic energy to peace, health and prosperity'. Further duties are to ensure that any assistance provided under its supervision is not used for military purposes. It sets safety standards and inspects to ensure that they are implemented. Later the Agency was given responsibilities for drawing up safeguards and verifying their use under the Non-Proliferation Treaty (NPT) of 1968. This Treaty, despite its imperfections in the way it protected the monopoly of the five pre-existing atomic weapon Powers, did appear to brake the further increase in the number of such Powers (though there are widely thought to be a further seven powers with illicit weapons). NPT was a response to worldwide public concern that this valuable source of power should not be misused to provide materials for the most powerful and destructive of modern weapons.

By 1990, IAEA had 113 members. It has organised safety assistance to the former Soviet Union and to Eastern and Central European countries. In general, these countries depend heavily on nuclear power for their electricity supplies and it has not been practicable to recommend closing down units. So the Agency operates several programmes focussed on each of the various types of Soviet-designed reactors[15]. In 1994 an apparent threat by North Korea to leave the IAEA and expel its inspectors was interpreted as being linked with diverting nuclear material for use as weapons. This led to an international crisis, which was resolved only when North Korea agreed to allow resumption of inspections by IAEA. An extension conference for the NPT is due in 1995.

In the year 1991 to 1992, use of nuclear power rose slightly in the USA, which accounts for almost a third of world usage, and in OECD Europe but fell sharply in the former Soviet Union. There were large relative increases in India and Netherlands, though in both cases from a small base. Usage in Japan, with over a tenth of all world nuclear power, continued to rise at the relatively vigorous rate of over 4%. Japan has a Long Term Programme, the work of its Atomic Energy Commission which sets future targets for its nuclear programme. Its utilities are planning further generating plants using advanced boiling water reactors (ABWR).[5] A prototype fast reactor (Monju) is starting up — it went critical early in 1994 — and a demonstration fast reactor is planned for construction 'in the early part of the next century'. Japan is reported to envisage commercialisation of fast reactors by the year 2030.

In France, following an intensive programme of development launched in 1974, nuclear electricity now represents 75% of total production of electricity in the country[4] and about a sixth of world generation by this means. In 1973 the oil crisis brought home to them the dangers of depending on imports for three quarters of their energy supply and led to both an energy conservation campaign and the intensive nuclear development programme. France exports electricity to Switzerland, Italy and Spain, and to the UK through a cable under the Channel. It is preparing plans for replacement of present nuclear plants when they become obsolescent. Électricité de France claims that it has been able to control costs of plants because of standardized technology, large industrial investments and continuity in its programme.

Austria does not generate any nuclear electricity. The country built a nuclear plant at Zwentendorf but following a national referendum passed a law forbidding nuclear generation and removed nearly all the machinery from the plant[7]. Meanwhile, Austria imports electricity from Germany and Switzerland, some of which is nuclear-generated. Sweden, currently taking over half its electricity from nuclear power plants, has also been left in a confusing situation due to public hostility to these sources indicated in a referendum in 1980. After much debate on economic and environmental consequences, the political parties agreed in 1991 that the dozen nuclear power plants would be closed down by 2010 but subject to conditions. The Riksdag (Parliament) confirmed[8] that '...the start of the phaseout shall be determined by the success of efforts to increase energy efficiency, by the availability of new, environmentally acceptable means of power production and by the goal of maintaining internationally competitive electricity rates'. A five year programme was approved for work intended to achieve these goals. However, Parliament has recently voted against setting 2010 as a specific legal date when nuclear power must be closed down[7], implying serious concern about the consequences.

Overall, it is now estimated[10] that nuclear power supplies 5% of the world's primary energy, or 17% of the world's electricity; at the end of 1992 there were 424 reactors in operation in a total of 29 countries.

Fusion

One of the key reasons for continuing interest in fast (fission) reactors and in nuclear fusion reactors was neatly summarised in this table of equivalences calculated in 1975[12] (the exact figures may vary a little depending on the assumptions made.)

To provide the equivalent of the world's annual electricity consumption in power stations with 40% efficiency on the basis of, for example, the 1975 figures we would have to burn;

- 1,700,000,000 tonnes of coal or
- 85,000 tonnes of uranium in conventional fission reactors or
- 1,000 tonnes of uranium in fast breeder reactors or
- 1,000 tonnes of lithium in D - T (deuterium - tritium) fusion reactors or
- 135 tonnes of lithium in D - D (deuterium - deuterium) fusion reactors.

The further attraction is that the fuels for nuclear fusion are plentiful and widely available throughout the world. Deuterium (heavy hydrogen) can be extracted from water and tritium (a further form of hydrogen) is manufactured in the fusion reactor from lithium, the lightest metal, which is plentiful in the earth's crust and easily obtainable. The process does not produce greenhouse gases nor gases involved in acid rain. No long-lived radioactive waste products are produced.

Nuclear fusion, seen by its enthusiasts as a vast new source of energy for human use, is the process taking place continuously in the sun and the stars. At unbelievably high temperatures, lighter nuclei fuse together to form heavier ones releasing immense amounts of energy. In a self-heated fusion reactor the gases deuterium and tritium will react to form helium and a neutron. Capturing the neutrons in a 'blanket' containing lithium leads to the formation of tritium.

The gaseous fuels must be heated to temperatures ranging up to no less than 200 million degrees Celsius, a temperature several times hotter than the centre of the sun. The gas under these conditions is in a state where it is known as a plasma. It must be isolated, prevented from touching any material surroundings by using magnetic fields. This system of confinement by magnetic fields is called a 'tokamak', which is an acronym from the Russian for 'toroidal magnetic chamber'. The EU, the USA, the former Soviet Union and more recently Japan have all had research programmes on nuclear fusion.

In 1991 the Director of the Joint European Torus (JET) announced that they had achieved fusion power — generating 2 MW in a pulse lasting for two seconds. (Torus is the shape of the main experimental vessel, sometimes described as a giant bun-shaped ring; the corresponding adjective is the word 'toroidal'). JET is a collaborative project of all the EU countries together with Switzerland and Sweden. The next objective is to design an experimental fusion reactor capable of generating more than 1000 MW of thermal power. This is planned as a world-wide collaboration involving the USA, Japan, Soviet Union and the EU in an International Thermonuclear Experimental Reactor known as ITER.

However, Dr R Hawley of Nuclear Electric is very sceptical about the prospects.[10] Fission and fusion, he points out, were discovered within a few years of each other. But fission power is now a 'maturing industry', while fusion has yet to demonstrate scientific feasibility. Though continuing progress has been made towards the dazzling vision of sustained reaction there are also practical engineering problems as well as the basic scientific one. Removing the heat from the fusion reaction involves handling very high rate of heat flow. In addition, 80% of the fusion energy appears in the form of fast neutrons. These are very damaging to structural materials so that, even with better materials, it is reckoned that the first wall of the structure would have to be replaced every few years. Finally, the fusion reactor would be many times the cost of a PWR of the same output — some estimates put it at ten times more expensive. Hawley's cynical conclusion is that 'when I was a young engineer, success was 40 years away. Forty years on it is clear that it is still 40 years off'.

Cold Fusion

In recent years there have also been scientific papers[17] reporting nuclear fusion under very much milder conditions, known as 'cold fusion'. The evidence essentially was that whilst passing an electric current through an electrochemical system, neutrons were detected and heat was generated far in excess of that corresponding to the electrical energy plus that due to possible chemical changes. These results have been disputed and other scientific teams in USA, Germany, the former Soviet Union, China and Britain have variously reported both positive and negative results. The area remains one of active research and debate[18].

UK Trends

In the UK, nuclear generation of electricity has grown from its virtual beginning before 1960 to an oil equivalent input amounting to 22% of total fuel input for electricity generation in 1992. The reason for this lumbering phraseology is that, in the UK statistics[14], nuclear contribution to electricity generation is expressed in oil (or sometimes coal) equivalent as the notional amount of fossil fuel needed to produce the same net output of electricity. The efficiency assumed is that of equivalent contemporary conventional steam power stations in the UK which averages 34%.

Research and development support for the UK nuclear power programme is provided by the UK Atomic Energy Authority, which also has responsibilities for public information. One division operates on a basis that is called 'quasi-commercial' under the trading name AEA Technology and provides scientific consultancy services on commercial terms worldwide. When the nationalised Central Electricity Generating Board (CEGB) was restructured in 1990 in preparation for privatising, its nuclear generating functions remained in public ownership as a result of strong indications from financial/commercial interests that the nuclear units would be impossible to sell. The CEGB was split into two generators based on fossil fuels , National Power and PowerGen, one nuclear-based generator called Nuclear Electric, and the National Grid Co in charge of the electricity transmission system and two pumped storage stations. The reasons given for the privatisation were that competition would reduce prices and improve services; prices have actually risen provoking documented complaints both from industry and from the National Consumer Council[29]. In Scotland, three new companies were formed; Scottish Power and Scottish Hydro-Electric were floated on the Stock Market in 1991 but not Scottish Nuclear Ltd.

For services covering the whole nuclear fuel cycle there is a further company called British Nuclear Fuels plc (BNFL) wholly owned by the government. Its major project is the thermal oxide reprocessing plant (THORP) built to reprocess spent fuel from British and overseas reactors using oxide fuels. After much dispute including legal actions, this was being brought into operation in mid-1994. BNFL also operates other waste management and effluent treatment facilities. Nuclear Electric has six Magnox stations, the earliest type of station using uranium metal as fuel and carbon dioxide cooling, and five Advanced Gas-cooled Reactor (AGR) stations using uranium oxide fuel slightly enriched in U235, also cooled by carbon dioxide but operating at a higher temperature. A PWR known as Sizewell B was completed in mid-1994; this uses uranium dioxide fuel with pressurised hot water containing boron both as moderator and to transfer heat to a second water system producing steam that drives the turbine. Sizewell B is claimed to have exceptionally advanced 'state-of-the-art' instrumentation and control systems. Scottish Nuclear operates two AGR stations. There are also three Magnox stations — Berkeley, Trawsfynydd and Hunterston A — that have been shut down and are in the process of being decommissioned.

A separate company was set up for dealing with solid low and intermediate-level waste, including research. It is called UK Nirex Ltd (Nuclear Industry Radioactive Waste Executive) and is owned jointly by Nuclear Electric plc, Scottish Nuclear Ltd, British Nuclear Fuels plc and AEA Technology. The government retains a special share in the company. BNFL is responsible for high-level and other liquid wastes among its other duties. As briefly mentioned earlier, the Royal Society has independently reviewed the work of NIREX and generally approved it.[25] However, it has warned that much of the science is close to the frontiers of knowledge so that the timetable needs to be flexible and much more research on the problems is needed.

An extra charge known as the 'fossil fuel levy' is applied as a surcharge on all electricity bills to cover the greater cost of generation from nuclear sources. In 1993 it raised 1.23 billion. 94% of this went to the nuclear power companies, most of it to Nuclear Electric. (Originally the purpose was to help cover the costs of closing down nuclear stations and extra costs in reprocessing and disposing of spent fuel). The balance goes to subsidise renewable sources of energy. The levy was originally due to be phased out by 1998 but the European Commission has raised no objections to the UK government extending the scheme for renewables up to the year 2014.[13]

AEA Technology and Nuclear Electric are both campaigning to be privatised. Following the example of the privatised generating companies in earlier years, both organisations have cut staff very severely in 1993 and 1994, Nuclear Electric blaming particularly price controls imposed by the Office of Electricity Regulation; this is the regulatory body set up under the Electricity Act 1989, which operates independently of the government.

Meanwhile, the government in mid-1994 set up two reviews. The Department of the Environment (DoE) will examine how to deal with nuclear waste and with decommissioning nuclear stations. Responding to the DoE Consultation Document the Royal Society has rejected the view that strategy for waste should be left to commercial operators.[25] There must be a coherent national policy formulated by the Government on the basis of thorough scientific research. In parallel the review by Department of Trade and Industry (DTI) 'will focus on the future prospects for nuclear power including, without commitment, its privatisation.' As noted above, without waiting for the results of these studies, and in the face of widespread criticism, the government has begun the process of shutting down the UK PFR. The critics commented that uranium will not always be cheap and plentiful, that the fast breeder is considered to be the reactor of the future, and that expert teams will be broken up. Social consequences in Caithness are likely to be severe. Overall a staff of 2500 will be cut to 500 with knock-on effects on the local economy, and a large community of scientists and engineers will be thrown out of work in the north of Scotland with little alternative employment available.

Exposure of the Public to Radiation

In response to public concern about risks due to radiation, the National Radiological Protection Board (NRPB) was established by the Radiological

Protection Act 1970. It is an independent body acting as a national point of authoritative reference. As an active research organization NRPB also provides information, advice and services in its area of expertise, including advising the government on the acceptability of recommendations by the International Commission on Radiological Protection (ICRP). Despite the frequently expressed concern on the dangers due to the nuclear industry, its last radiation review[13] dealing with exposure in 1991 concluded that:—

- occupational doses in the nuclear industry continue to decline steadily and are now about half of the level at the last review (covering 1987);
- doses from nuclear discharges are trivial for most people and for the most highly exposed groups well within the annual dose limit.

The dominant average source of exposure for the public, providing half the average radiation dose, has no connection with the nuclear industry but is due to radon, a naturally occurring radioactive gas (and its decay products) leaking into houses. Radiation from all natural sources — including cosmic rays, gamma rays from rocks and building material, food and drink, as well as radon — totals about 85% of the average dose. Exposure of patients to X-rays is next in importance. These are in line with the conclusions of the UN Scientific Committee on the Effects of Atomic Radiation.

Radiation Exposure of the UK Population - 1993 Review

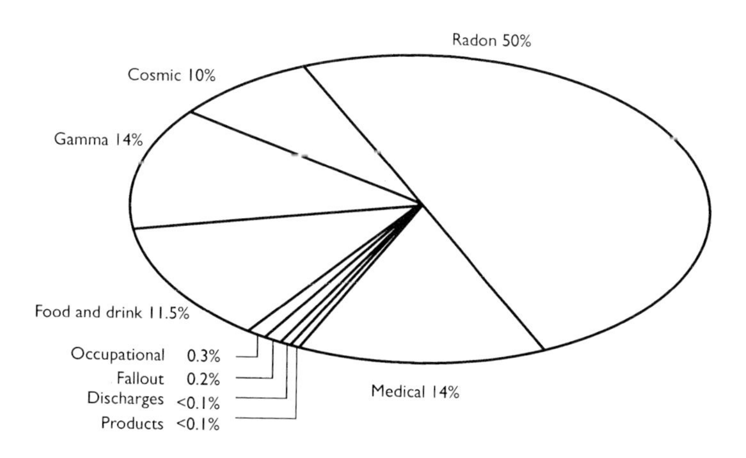

Source: NRPB.

Occupational Exposure of Radiation Workers

In 1976 NRPB established the National Registry for Radiation Workers. This gathers information on their radiation doses and analyses causes of deaths. The first analysis[21], for 1986 and published in 1992, covered over 95,000 people working in the Atomic Weapons Establishment, British Nuclear Fuels, the Defence Radiological Service, Nuclear Electric and the UK Atomic Energy Authority. Over 6,000 had died by the end of the study period. The Standardised Mortality Ratios (death rates calculated to allow for the distribution of ages) for all causes and for all cancers combined were well below 100, that is below the rates for the general population. Cautiously the researchers commented that this 'Healthy Worker Effect' is a common observation in occupational epidemiology. They also added that they planned further analyses allowing for a larger study population and longer follow-up. In the first analysis they also found mortality from most types of cancer was below that in the general population. An exception was thyroid cancer where the death rate was significantly above that in the general population. Other effects were discovered but were reported with a note that they did not reach statistical significance, that is to say they might have arisen by chance.

Concerns about leukaemia and other cancers in the children of male workers at Sellafield led to two studies[30] by the Health and Safety Executive. These concluded that there was no statistically significant association (as had been suggested at one stage) between the incidence of leukaemia and non-Hodgkins lymphoma (NHL) in children, and the father's external radiation dose. This work finally laid to rest anxieties prompted by an academic researcher's suggestion that there might be such an association linked to workplace exposure.

References

1 *Development of atomic energy. Chronology of events 1939-1978* Harwell: UKAEA Authority Historian's Office, Atomic Energy Research Establishment, 1979

2 *BP Statistical Review of World Energy 1993* London: British Petroleum Co. plc., 1993

3 *Whitaker's Almanac 1992* 124th ed., London: J Whitaker and Sons Ltd., 1992

4 R Carle 'French energy policy' paper delivered to: 'Technology in the third millennium: energy' discussion meeting of the Royal Society and Royal Academy of Engineering, London, March 1994. Published in D Rooke, I Fells, J Horlock *Energy for the future* London: E and FN Spon for the Royal Society, 1995 Technology In The Third Millennium no. 6 ISBN 0419200509

5 'World-wide industry eyes the expanding market' *Atom* no. 433 (Mar/Apr 1994) pp.14-18

6 'More power to the people of Southern China and Hong Kong' *Atom* no. 433 (Mar/Apr 1994) pp.19-21

7 *Power in Europe* no. 175 (Jun 3 1994)

8 *Energy and energy policy in Sweden. Fact sheets on Sweden* Stockholm. 1991 FS 37 pp.P08 ISSN 1101-6124

9 Educational and informational booklets issued by the UK Atomic Energy Authority and the British Nuclear Forum, such as 'Glossary of Atomic Terms', 'Nuclear Power Reactors', 'Nuclear Power and Radioactive Waste' and others

10 R Hawley 'Nuclear power - energy for the 21st Century. Meeting the challenge' paper delivered to: 'Technology in the third millennium: energy' discussion meeting of the Royal Society and Royal Academy of Engineering, London, March 1994. Published in D Rooke, I Fells, J Horlock *Energy for the future* London: E and FN Spon for the Royal Society, 1995 Technology In The Third Millennium no. 6 ISBN 0419200509

11 Tony Broomfield 'PFR builds firm foundations for fast reactors' *Atom* no. 433 (Mar/Apr 1994) pp.22-26

12 W Hafele *Fusion and fast breeder reactors* Laxenburg, Austria: IIASA RR-77-8 quoted in *Nuclear fusion. Energy for the 21st Century* Oxford: JET, CR 87.129

13 J S Hughes and M C O'Riordan *Radiation exposure of the UK population - 1993 review* Didcot, Oxon.: National Radiological Protection Board/HMSO, 1993 NRPB-R263 ISBN 0859513645

14 *Digest of United Kingdom Energy Statistics 1993* London: HMSO, 1993

15 *International assistance to upgrade the safety of Soviet-designed nuclear power plants. Selected activities in Eastern and Central Europe and the countries of the former Soviet Union* Vienna: International Atomic Energy Agency, 1993

16 The confusion continues. According to the *Guardian* (8 Jul 1994) the Ministry of Defence admitted that plutonium from Britain's civil nuclear programme had been used in nuclear weapons testing in the USA. Yet on 28 June 1993, Lady Chalker had stated for the government 'We would only allow the export of plutonium from processing if we had received satisfactory assurances that it would only be used for peaceful non-explosive purposes'

17 For instance Martin Fleischmann 'The present status of research in cold fusion' in: section 33 'Science of Cold Fusion' *Conf. Proc. Italian Physical Society*, 1991 pp.475-87

18 Other papers at the conference noted in Ref 17. For instance Heinz Gerischer 'Is cold fusion a reality? The impressions of a critical observer' in: section 33 'Science of Cold Fusion' *Conf. Proc. Italian Physical Society* 1991 pp.465-74

19 *Nuclear power and radioactive waste* London: UK Atomic Energy Authority, 1991

20 For example *Clearing the air. The nuclear answer to the energy question* London: British Nuclear Forum, 1993

21 G M Kendall et al. *First analysis of the National Registry for Radiation Workers. Occupational exposure to ionising radiation and mortality* Didcot, Oxon.: National Radiological Protection Board, 1992 NRPB-R251 and
G M Kendall et al. 'Mortality and occupational exposure to radiation; first analysis of the National Registry for Radiation Workers' *British Medical Journal* (1992) pp.220-225

22 Terence Lee 'Risks and perceptions' *Times Higher Education Supplement* (8 May 1987)

23 *Risk assessment. A study group report* London: The Royal Society, 1983 and *Risk: analysis, perception and management* London: The Royal Society, 1992

24 *The management and disposition of excess weapons plutonium* Washington: US Academy of Sciences. Quoted in 'Beating swords into plowshares' *Atom* no. 434 (Jun/Jul 1994) pp.39,40

25 *Disposal of radioactive wastes in deep repositories* London: The Royal Society, Nov 1994 ISBN 0854034935 and *Radioactive waste management policy. Response to DoE consultation document* London: The Royal Society

26 *The global uranium market. Supply and demand 1992-2010* London: The Uranium Institute, 1994 ISBN 0946777276

27 'First German reactor disassembled' *Atom (*Aug/Sept 1994) p.435

28 Brochures and pamphlets of Friends of the Earth, 26-28 Underwood St, London N1 7JQ. And Peter Roche 'The United Kingdom - world class proliferator' *CND Today* (London, Greenpeace) (Spring 1993)

29 *Electricity distribution. Price control, reliability and customer services: Response to offer* London: National Consumer Council, Feb 1994 PD 01/E/94

30 *HSE investigation of leukaemia and other cancers in the children of male workers at Sellafield - Review of results published in October 1993* HSE Books, 1994 ISBN 0717608336

intentionally left blank

Chapter 6

Renewables

Contents

- What are Renewables?
  - Municipal Solid Waste
  - Dry Wastes as a Fuel
  - Wet Wastes as a Fuel
  - Crops as Fuel
  - Wind Energy
  - Tidal Power
  - Hydro Power
  - Wave Energy
  - Geothermal Hot Dry Rock
  - Geothermal Aquifers
  - Photovoltaics
  - Photoconversion: Electrochemical Photovoltaic Cells
  - Photoconversion: Photobiological Processes
  - Active Solar Heating
  - Passive Solar Design
  - Ocean Thermal Energy Conversion
  - Solar Ponds
  - Improved Solar Systems
  - Tidal Current Turbine
- Environmental Aspects
- World Trends
- The Developing Countries
- European Trends
- UK Trends
- References

Renewables

"When the sun shineth, make hay"

John Heywood. *Proverbs*, pt1, Ch 2.

All humans over a very long period depended on traditional biomass for fuel — crop residues, fuelwood, dung — as well as wind, sun and water power as energy sources.

Somewhere ahead of us there is a very sunny future. Nobel Laureate, Professor Lord Porter envisages[1] a second green revolution 'when it becomes possible for us to design photosynthesis, and indeed evolution itself, so that it makes the products we want efficiently, economically, and with the minimum need for energy or further processing'. The renewable products of photosynthesis will then take the place of fossil fuel reserves.

Most of this chapter deals with the time span between these two periods.

What are Renewables?

Though fossil fuels are generally more compact, cheaper by conventional costing, and much more controllable as sources of energy than the earlier renewable sources — which is why the fossil fuels took over in the industrial world — there has been growing interest in applying modern science and technology to using renewable sources. In the earlier stages of this movement, the spur was the risk of exhausting fuel reserves. Today the emphasis has switched rather to concern about environmental damage — greenhouse gases, acid rain, health-threatening pollution — due to using fossil fuels. But a further asset of renewable sources is seen in the fact that they increase diversity and therefore security of energy supplies. What then are these 'renewables?'

In practice the term is used to refer to a wide range of naturally occurring energy sources but also to others that might more accurately be described as replenishables including wastes from various activities. The UK Renewable Energy Advisory Group (REAG) — a group of experts set up to advise the government on a national strategy — defined renewables[2] as 'those energy flows that occur naturally and repeatedly in the environment and can be harnessed for human benefit. The ultimate sources of most of this energy are the Sun, gravity and the Earth's rotation.'

An extensive investigation of the possible future contribution of renewables by the World Energy Council (WEC) concentrated[3] on what it defined as the new renewable forms; they were solar, wind, geothermal, modern biomass, ocean (including tidal, wave, ocean thermal and salt gradient), and small hydro. REAG's formal list, with its explanations, comprised these 'technology modules':—

Municipal Solid Waste

(MSW) Wastes produced by households, industry and commerce. Energy can be recovered by combustion directly or after reclaiming recyclables such as metals and glass, or by landfilling and using the methane-rich gas.

Dry Wastes as a Fuel

Industrial, agricultural and forestry residues that can be burnt as fuel or gasified and burnt. A recently developed fermentation process alternatively can convert agricultural waste into pure alcohol for fuel.

Wet Wastes as a Fuel

Sewage sludge, industrial effluents and farm slurries that can be bacterially fermented in the absence of oxygen. Gas produced is typically a mixture of 65% methane and 35% carbon dioxide, generally known as biogas and used for direct heating or for generating electricity.

Crops as Fuel

The crops may range from wood to sea algae and are grown deliberately as fuel as distinct from collecting naturally-occurring combustible matter. There are known energy crop programmes in Brazil, Sweden, elsewhere in Europe and North America as well as the UK. Trees are carefully coppiced so that they re-grow. Cereals may be grown to produce ethanol, or oilseed crops to produce an oil-based substitute for diesel.

Wind Energy

Modern developments in using this source — used in more primitive ways for over 2000 years — make it one of the most promising for generating electricity. The machines may be designed with a horizontal axis and propeller-like blades (though the detailed design is different from that of propellers) or, less successfully so far, with a vertical axis. The latter have a variety of designs of devices for catching the wind. Groups of horizontal axis machines (wind turbines) may be arrayed in 'wind farms' generating electricity.

Tidal Power

Tides are caused by the gravitational attraction of the moon and the sun raising and lowering the surface of the oceans. Energy can be extracted from the amplitudes of the tides, which are increased in estuaries. This too is a long-established source of energy. Tidal mills have worked on the coasts of France, Spain and Britain since 1100 AD. The first and largest modern plant (240 MW) is at La Rance in France where it has operated successfully for over 25 years with axial flow generators.

Hydro Power

Energy available from water flowing in a river or a pipe can also be tapped and systems of this kind have been used for some 2000 years. Hydro power is very widely applied today. It is the only form of renewable energy system used on a sufficient scale — generally in the form of large plants — to have entered for many years into world statistics for commercial energy[4]. There is also growing interest in small-scale hydro schemes particularly in developing countries but also in the industrial countries.

Wave Energy

The ocean waves are caused by the action of the winds as they blow across the surface of the sea and transfer energy to it. Though Salter's 'nodding ducks' were the devices for reaping some of this energy that most strongly captured the public imagination, over 300 design concepts were considered in the UK alone . Yet only experimental plants are currently in action. The western coastline of Europe is, however, estimated to have one of the largest wave energy resources in the world from waves travelling across the Atlantic.

Geothermal Hot Dry Rock

Below the Earth's surface is a large amount of heat stored in hot dry rocks (HDR) — 'geothermal' meaning related to the internal heat of the earth. This might be tapped by drilling two holes from the surface, pumping water down one of them, through the fissures in the rock to the other hole, where the water would return as superheated water or steam. So far, attempts at HDR technology in mainland Europe and in the UK have not been successful.

Geothermal Aquifers

When water is held in a layer of porous rock the structure is known as an aquifer. In many parts of the world — Iceland, Hungary, USA, France — geothermal aquifers are accessible from the surface by drilling. The hot water or steam is passed through a heat exchanger on the surface and the heat is used. In the UK so far only one site, at Southampton, has proved to be commercially usable; it is the basis of a local district heating scheme.

Photovoltaics

As the name indicates, these are devices based on materials that generate direct current electrical power when they are exposed to light. They apply semi-conductors, are expensive and have so far been used in space vehicles, in equipment remote from an electric grid, and in some consumer goods such as calculators and watches. Photovoltaics (PV) are an active area of research and prices have been falling. Commercial prospects for the future are thought to be promising.

Photoconversion: Electrochemical Photovoltaic Cells

Electrochemical Photovoltaic Cells (ECPV). These cells, converting sunlight directly into electrical power by electrochemical methods, are described as 'functional equivalents of conventional solid-state semiconductor photovoltaic cells' where the light-reactive centre is on the electrode surface. They are less developed than the PV cells.

Photoconversion: Photobiological Processes

Photobiological Processes. Systems of this kind include some based on anaerobic photosynthetic bacteria converting solar energy into hydrogen derived ultimately from water. (Anaerobic means growing without air). There are several other lines of research on photobiological processes as indicated in the opening quotation from Professor George Porter[1].

Active Solar Heating

This is probably the best-known application of solar energy. Solar heat collectors may consist of black-surfaced flat plates mounted behind clear glass sheets or an assembly of evacuated tubes. Water with anti-freeze or sometimes other liquids circulate over the collectors and transfer the solar heat in heat exchangers so that it preheats water in a conventional domestic hot water system. They are widely used in many of the sunnier countries such as Israel, and in more limited numbers even in the UK. Variants are applied for heating swimming pools. Experimental installations in sunny areas have used large parabolic mirrors to concentrate the heat to generate steam in boilers for power use, and even to attain much higher temperatures capable of melting metals.

Passive Solar Design (PSD)

This approach aims to maximise free solar gains to buildings by the design of the building itself. It may include the use of conservatories and roof space collectors. For warming the building this means arranging the glazing to take full advantage of solar heat, while avoiding losing heat from windows in the shade. Minimising the need for artificial lighting is also part of the design objective. Less obvious is using this approach in such a way that it assists cooling in hot weather by exploiting solar heated air to increase natural convection and promote extra ventilation. PSD is most effective when buildings applying these principles are grouped to avoid shading and even achieve protection from prevailing winds but using PSD means asymmetric layout of windows and to some observers the house may look odd.

Some experimental houses have been built incorporating several renewable energy systems. In Freiburg, Germany, (obviously temperate zone, not tropics) the Fraunhofer Institute for Solar Energy Systems has built a house where the entire energy demand for heating, domestic hot water, electricity and cooking is supplied solely by solar energy. The house is heavily insulated (using transparent insulation), with all occupied rooms orientated towards the south, while corridors and the staircase face north. Energy is stored by electrolysing water during summer by electricity from the PV generator and storing the hydrogen and oxygen under pressure for use in winter. A family has lived in the house for at least two years[9].

Other methods have also been investigated, such as the following; -

Ocean Thermal Energy Conversion (OTEC)

The temperature difference between the surface of tropical seas and deep water has been used in experimental power plants. Among others the Netherlands and Indonesia have looked at the possibilities of a site off the north west coast of Bali where the surface is at 30C, while, at 500 m depth the sea temperature is 8C. In this design[6], the hotter water would be pumped to the plant and by means of a heat exchanger vaporize ammonia; this would drive a turbine linked to an electric generator, then be condensed and cooled in a further heat exchanger by the cold water, and recirculated. A status report in 1993 considered that after extensive testing in the US and Japan the system was ready for commercialization[18]; a power producing experiment was reported to be under way in Hawaii.

Solar Ponds

In areas generally of strong insolation, shallow ponds with a lower layer of concentrated salt solution and upper layer of water show a temperature difference between the layers. The solar heat is much more strongly absorbed by the salt solution layer and its greater density minimises convective mixing. Systems essentially of this kind have been investigated in many tropical or semi-tropical areas[19].

Both this system and OTEC call for equipment capable of using small temperature differences to drive a heat engine that can then be applied to turn a generator. Several of the solar energy devices have also been applied experimentally to driving cooling systems — a particularly favourable aplication since the maximum natural power is then available at the time of maximum need for the service.

Improved Solar Systems

Solar heat has of course been used for many millennia for drying — from washed clothes to food for preservation. Researchers, for example at Hohenheim University in Germany, have developed improved systems of solar drying claimed[7] to dry tropical fruit and vegetables in ways that minimise weight loss and maintain quality, and have demonstrated their use in Crete. There are also solar stills for producing fresh water from brackish water.

Tidal Current Turbine

The Intermediate Technology group has designed[8] a tidal current turbine to tap the energy of tidal currents without using any civil works (as needed by tidal barrages). As a wind turbine is driven by the wind, so this submerged and anchored tidal turbine is turned by the tidal currents. A prototype is being installed near Fort William, Scotland. Its first application will be to generate electricity to charge a battery for a buoy tethered above the turbine.

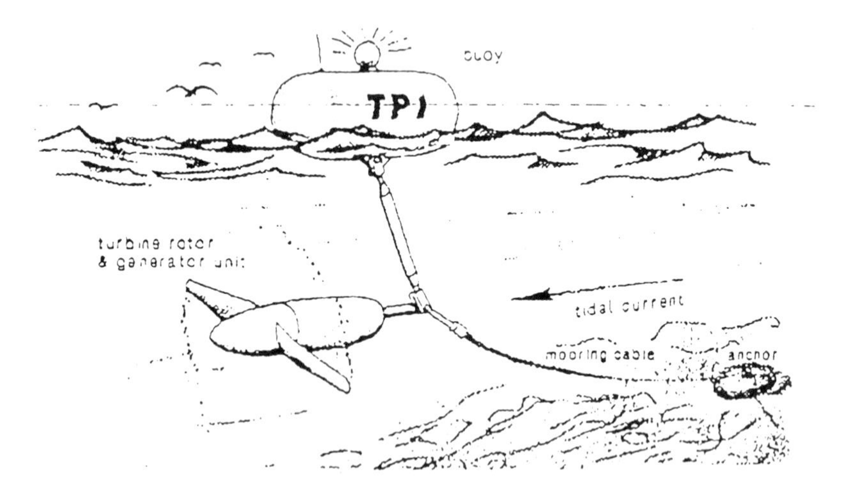

Artist's impression of a tidal current turbine, a concept developed by IT Power Ltd, Scottish Nuclear Ltd and the National Engineering Laboratory
Source: IT Power News, No. 10

Environmental Aspects

Above we mentioned briefly the concerns about pollution — notably acid rain damaging human health, vegetation, fisheries and buildings — and generation of greenhouse gases, as major spurs to current interest in developing renewable sources of energy. Overall these sources are more inherently benign than energy from fossil fuels. For example, the World Bank, anxious to establish its environmental *bona fides* in the face of widespread criticism, has started to include renewable energy operations in its Energy Policy programme[5]. In 1992 it supported its first operation of this kind — the India Renewable Resources Development Project which will finance forty-five mini hydroelectric plants. It also supports the Asia Alternative Energy Unit aiming to integrate renewable energy components with energy-efficient services in the region; here, the Bank's first projects were in Indonesia and Sri Lanka.

Yet these more benign sources also need specific studies to assess their environmental impacts. Most of the technologies listed above do not result in polluting emissions. Those that do not involve combustion produce no gaseous emissions and therefore no greenhouse gases nor gases contributing to acid rain. Where the renewables are fuels they are considered carbon-neutral since the carbon has been fixed in recent times and forms part of the current active carbon cycle. When the biofuels are burnt they produce no more emissions of CO_2 than they would in nature. However, on a world scale, against this must be set the fuelwood crisis in countries where people are gathering wood without re-planting. UN agencies (UN Environment Programme and the Food and Agriculture Organisation) after reporting that countries notably in Africa have been extensively stripped of their tree cover, have run campaigns for re-planting trees. In addition to renewing their primary product which may be fuel or cash crops — the trees also protect watersheds, conserve soil, provide habitat for wildlife, store genetic diversity and help to regulate climate.

Compared with fossil fuels with their polluting emissions of sulphur dioxide (SO2) and oxides of nitrogen (NOx) the fuels from current biological sources yield negligible amounts of these contributors to acid rain. Renewables are also considered to have favourable effects in respect of methane — an important greenhouse gas — in the atmosphere. The extraction of coal, oil and gas results in leakage of methane; further methane escapes into the atmosphere from gas leaks and from older landfill sites that have not been properly sealed and equipped. So, displacing the fossil fuels that would otherwise be used reduces the accompanying escapes of methane, while burning the landfill gas directly destroys this source yielding CO_2 - undesirable, but having less greenhouse effect than the very potent methane. (Molecule for molecule, methane is about twenty one times as effective as CO_2 as a greenhouse gas[14]. It does also arise in ways that are much less amenable to reduction by human efforts. They include emissions from wetlands, rice paddies, ruminant livestock, termites, oceans and freshwaters. But the human-influenced contributions can be greatly reduced).

Wind energy sometimes raises objections on the grounds of irritating noise and visual intrusion but its advocates claim[23] that people living near wind farms find the reality more acceptable than the anticipation, so that there are few complaints after wind farms have started operating. This view is based largely on a formal investigation[25] of before- and-after attitudes of a group in Cornwall living near a wind farm site, compared with those of a control group providing a comparison. Further support for wind energy came — rather surprisingly — from the House of Commons Welsh Affairs Committee[26]. After consulting a wide range of interests it concluded that wind power can be developed 'without causing unacceptable visual intrusion, without undue annoyance to local communities and without destroying valued landscapes' . In effect they rejected the objections of the Council for the Protection of Rural Wales which had called for a moratorium on wind farms.

Hydro and barrage schemes may involve flooding inhabited areas and displacing farmers from their lands. A very large-scale example of this kind was the Three Gorges dam in China creating a reservoir 600 km long and displacing around one million people from their homes. The plans caused controversy in the country and approval had been delayed for years[15] because of worries about the environmental consequences as well as the cost of the project. There were somewhat similar controversies in the 1950s and early 1960s about the Volta River Project in Ghana. Though this provided electrical generating capacity of 640 MW in a country previously consuming only 120 MW, together with new opportunities for fishing and for communication, its construction involved flooding over 3% of the country and resettling 80,000 people. An assessment in 1980[27] also referred to damaging side-effects in the spread of schistosomiasis and onchocerciasis (river blindness). Schemes of this type also alter water and estuarine habitats affecting wildlife so that modern projects have to take into account ways of minimising these effects.

In respect of ocean thermal energy schemes, the World Energy Council has warned that despite the enormous reserves of energy in tropical oceans, there may still be 'question marks' over OTEC because of ecological and climatic impacts.

World Trends

'Every effort should be made to develop the potential for renewable energy which could form the foundation of the global energy structure during the twenty-first century' urged the 1987 report of the World Commission on Environment and Development[13]. Generally, energy experts though hopeful have made more modest forecasts. But Dr Kevin Brown, Director of the UK government's Energy Technology Support Unit (ETSU) has commented[10] that 'predicting the renewables' future has become almost an industry in itself, with considerable intellectual resources and powerful computer models dedicated to the pursuit of authoritative answers.' With this somewhat cynical reservation we can note the results of two recent major studies. Both look forward to large expansion in the contribution to energy supply by renewables.

By the year 2050, according to the estimates of a UN commissioned study[16] renewables could provide 60% of world electricity and 40% of direct fuel use. This 'scenario' assumes a world economy in expansion and adequate support so that renewable energy technologies can meet much of the growing demand at prices lower than those usually forecast for conventional energy. The report reviews three major examples of renewables development to highlight the lessons that can be learnt. In Brazil the government has promoted the use of ethanol from sugar cane as a transport fuel — sometimes called a biofuel — but the scheme was hit by reduced oil prices. California has developed major supplies of both wind generated and solar thermal generated electricity but companies have gone bankrupt as subsidies have been withdrawn. Denmark's wind energy programme is quoted as a genuine success story with strong support from the government though it has withdrawn its direct subsidy. Denmark now produces 2.5% of its electricity from wind power. (In 1990, the Association of Danish Windmill Manufacturers, FDV, claimed they had provided 900 MW wind power capacity in 28 different countries, including nearly 600 MW to USA).

Conclusions are that renewables have huge potential but need commitment by governments, developers and consumers if this potential is to develop as a viable energy option.

More or less in parallel, the World Energy Council had been carrying out its own study which reached much more restrained conclusions[17]. Even with clear and widespread public policy support, the 'new' renewables would take many decades to develop and diffuse to the point where they significantly substitute for fossil fuels in the view of the WEC. Though in 1990 nearly 18% of the world's primary energy came from renewables, over 98% of this was from total biomass and hydro. If current policies continued the contribution of renewables could relatively easily rise to 21% by the year 2020, or with strong international support at government level it might even reach 30% with much of the impact felt in developing countries.

Price of Electricity Produced by Danish Wind Turbines in 1980-89 and a Forecast for 1990-95

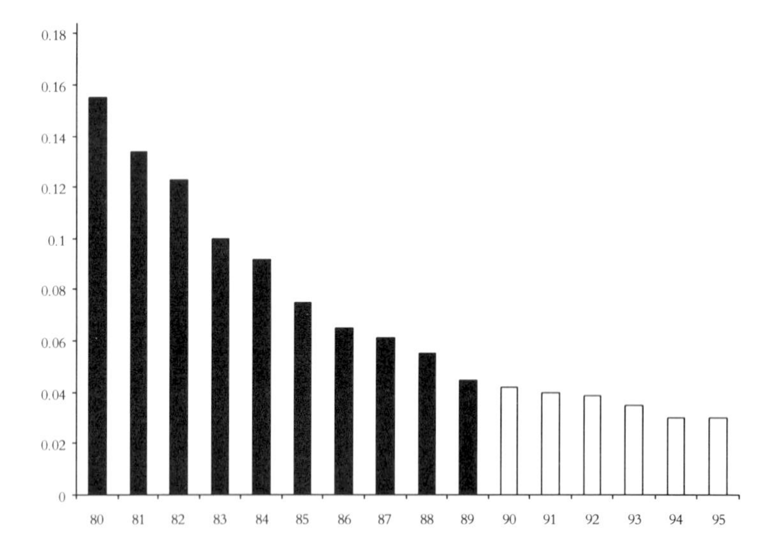

Source: from leaflet of Association of Danish Windmill Manufacturers/FDV

The WEC was in general opposed to government subsidies yet saw it as essential to promote accelerated development of new renewable energy. This should include subsidies through the research and development stages to the demonstration stage but should not go beyond.

The Developing Countries

Renewables are seen as having particular virtue in aiding economic growth in developing countries because they are based on diverse and distributed sources[10].

They are therefore readily accessible to regions beyond the economic reach of centralised power generation and distribution systems.

In tune with this line of thought is a Global Photovoltaic Action Plan prepared by the Commission of the European Union. Its aim is to initiate global action to supply electricity to more than a thousand million people who are without electricity today. The Plan is based on a distributed and 'decentralised electricity system of very low capacity for minimum basic needs...the only realistic system which could be implemented in a foreseeable time frame of 20 to 25 years.' Anticipating the criticism of expense, Dr Wolfgang Palz in presenting it[11] made two telling points. A conventional electric grid development would be too costly and time-consuming, and 'the cost of the project will be less than 1% of global military expenses.'

Further contesting the point that photovoltaics is too expensive, Palz observes that PV even at 50 cents per kWh is today the cheapest source of electricity in large parts of the world's rural areas. The importance of the project lies first in the local human sense noted earlier for improving infant mortality rates, life expectancy and literacy; but what may be more effective in gaining political approval and funding is the further argument that improving amenities in villages will have a dramatic effect in fighting the 'exploding demography' in the Third World, social unrest and migration. So, important features of the Plan are that it must

- enrich social life in the village, particularly in the evenings;
- involve young people, stimulate agricultural and cottage industry and create jobs;
- include and promote existing PV industry in the third world.

A model scheme for a village of 600 people in 40 families of 15 each provides for water disinfection, a health centre, dental care, family lighting, insect fighting, a refrigerator for the village shop and some other services — allowing 10 Watts peak capacity of PV per head. The receiving country would have to provide some finance, while the local villagers would contribute modest finance and assemble the complete systems locally.

Photovoltaics already contribute greatly to the Cold Chain programme of the World Health Organisation (WHO). This is a scheme to ensure that vaccines needing storage at low temperatures are maintained active from the manufacturer through all stages of transport and storage to the health centre where they are used.

The earlier stages are likely to have normal mains power supplies but the final stages, for example in isolated African villages, often depend on solar energy to supply power for the refrigerator. WHO has issued guidance and product information books on photovoltaic refrigerator systems for its Expanded Programme on Immunization. There are also tables of maximum storage times and temperatures for named vaccines at different stages along the Cold Chain.

On a more modest scale both practical guidance through working projects, and advisory publications, are available from non-governmental bodies such as the Intermediate Technology Group. Their aid includes publishing guides to energy in rural development programmes and catalogues of equipment covering a wide range of renewable sources but also suitable internal combustion engines for these applications. In the most recent power manual in this series[12] are also discussions of two important practical points.

The first is examining the circumstances in which renewables offer practical or economic advantages over more conventional power sources such as the nearest point on an electric grid, or using an internal combustion engine. The second group of issues are the socio-economic ones. Will the renewable energy system proposed be locally acceptable — in terms of cost, benefit, culture, convenience, skills, safety and the social question of who will be in control? To function properly, the energy technology will need skills for maintenance and operation, structures for management and support, and probably a system of credit. So these are the essential areas of study alongside the technical ones involved in choosing equipment.

Trends in the European Union

In 1993 the Council of Ministers of the EU agreed 'to endeavour to contribute in their energy policies to the limitation of carbon dioxide emissions by taking account of the Community's indicative objectives relating to the renewable energy sources...'. Though the decision[20] is evidently hedged around, it is clearly very much better than not 'endeavouring' at all. The specific objectives were these: —

To reduce carbon dioxide emissions by 180 million tonnes by the year 2005 by

(a) increasing the contribution of renewable energy sources to covering total energy demand from less than 4% to 8% in 2005. Production of renewable energy sources should rise from nearly 43 million tonnes oil equivalent (mtoe) to about 109 mtoe in 2005.

(b) trebling the production of electricity from renewable energy sources (excluding large hydro-electric power stations) from about 8000 MegaWatt (8GW) capacity to 27 GW.

(c) securing for biofuels 5% of total fuel consumption by motor vehicles. This means producing 11 mtoe of biofuels in 2005.

In addition, as we noted in the opening chapter, the EU announced early in 1994 for the first time, figures for the current contribution by renewable energy (RE). Production from such sources represented 6.7% of total primary energy production but RE consumption was 3.5% of gross inland energy consumption. (EU production of primary energy is only about half its energy consumption). Hydropower formed 8.2% of total EU electricity production, France being the biggest producer.

In support of the Council main decision it also agreed to allot ECU 40 million for the period 1993 to 1997 to be used for specific actions in support. They were: —

(a) technical evaluations for defining specifications;

(b) training and information and a wide range of pilot projects;

(c) creating an information network;

(d) assessing use of biomass for energy purposes.

UK Trends

'Renewable sources can and should make a significant contribution to future energy supply in the UK' reported the Renewable Energy Advisory Group (REAG)[2]. Their recommendations strongly supported government intervention, favouring a partnership with industry to help develop a supply industry and establish a self-sustaining market by the year 2005. Though there was a wide range of possible sizes of the contribution by renewables the 'plausible figure' for the contribution in 2025 under severe pressures of need and economics would be around 20% of the 1991 electricity supply. Of the specific technologies, REAG saw hydro large and small-scale, and solar water heating as 'close to technical maturity'. Quite modest government expenditure should remove outstanding queries on others including:—

- energy from wastes;
- energy from crops;
- horizontal axis wind turbines;
- passive solar gain;
- tidal power environmental effects.

As part of the study of removing barriers and constraints, a number of administrative issues were considered. Both REAG and earlier a Parliamentary Energy Select Committee wanted more attention to the external costs of energy sources including environmental costs. If prices to the consumer reflected these external costs, REAG indicated, it would greatly influence the competitiveness of renewables.

REAG also wanted to split the Non-Fossil Fuel Obligation (NFFO). Under this scheme, the Secretary of State can make orders for Public Electricity Suppliers to secure specified amounts of electricity generation capacity from specified renewable energy sources. This electricity is more expensive so the suppliers charge a levy to customers — in 1993/94 an extra 10% on their bills thus raising £1.233 billion. Only a small proportion of this went to renewables — though government press statements and publications imply very much more. An overwhelming 94% of the total[24] went to support nuclear energy activities (REAG pointed out that the primary purpose of the NFFO when it was introduced in the Electricity Act 1989 was to support the nuclear industry). The Group recommended splitting off the portion for renewables and calling it the Renewable Energy Obligation (REO) and it should be broadened to cover heat-producing applications as well as promoting the generation of electricity.

After considering this and other reports[21] the government published its own policy[22]. This is 'to stimulate the development of new and renewable energy sources wherever they have prospects of being economically attractive and environmentally acceptable in order to contribute to:

- diverse, secure and sustainable energy supplies;
- reduction in the emission of pollutants;
- encouragement of internationally competitive industries.'

Their programme will work towards 1,500 MW of Declared Net Capacity (DNC) from renewable sources for the UK by 2000. The present contribution of renewables is estimated to be less than 1% of primary energy and about 2% of the 1992 electricity supply. The main component is 1200 MW of hydro-capacity in Scotland mainly from large-scale schemes.

Renewable Energy Utilisation 1992

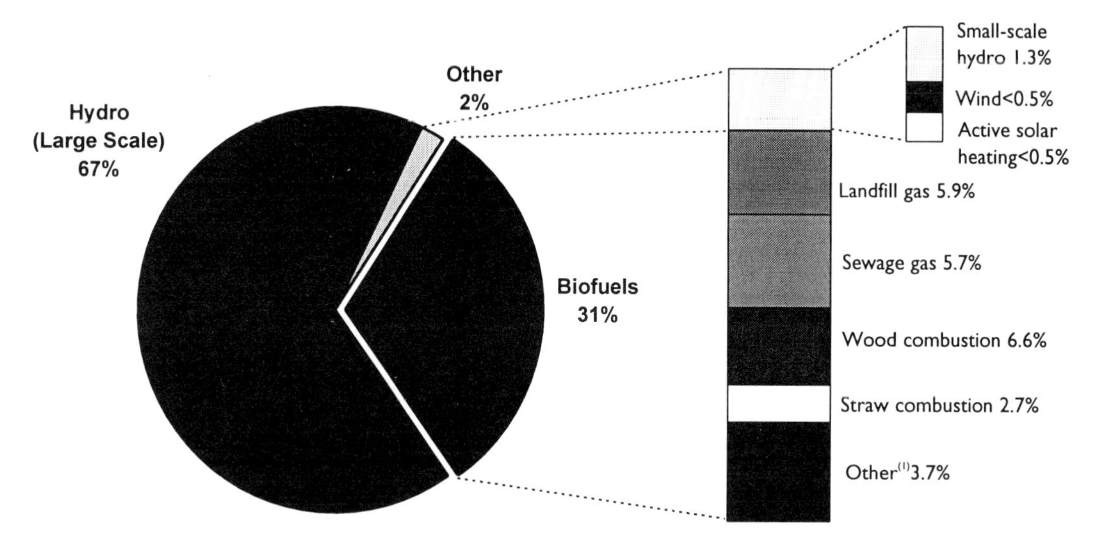

Total renewables used = 2.48 million tonnes of oil equivalent

(1) "Other" includes farm waste digestion and chicken litter, industrial and hospital waste combustion.
N.B. This figure excludes unplanned use of passive solar energy, estimated to be about 11.8 million tonnes of oil equivalent annually. Reference: Digest of UK energy statistics, 1993, HMSO.

Source. Energy Paper Number 62, p 5, HMSO

Other technologies are at widely varying stages of development. Thermal solar — for reasons evident to the general public too — is considered inappropriate for the UK.

Current Status in the UK of the New and Renewable Energy Technologies

Current Status in the UK of the New and Renewable Energy Technologies

	Inappropraite for the UK	Research	Development	Demonstration	Commercially Available	Established Market
Hydro Large Scale					xxxxxxxxxxxxxxxxxxxxxxx	
Simple Passive Solar				xxxxxxxxxxxxxxxxxxxxxxx		
Hydro Small Scale					xxxxxxxx	
Landfill Gas				xxxxxxxxxxxxxxxxxxxxxxx		
Active Solar					xxxxxxxx	
Onshore Wind Energy			xxxxxxxxxxxxxxxxxxxxxxxxxxxxxxxxxxx			
Specialised Industrial Wastes			xxxxxxxxxxxxxxxxxxxxxxxxxxxxxxxxxxx			
Municipal and Industrial Wastes			xxxxxxxxxxxxxxxxxxxxxxxxxxxxxxxxxxx			
Advanced Passive Solar			xxxxxxxxxxxxxxxxxxxxxxxxxxxxx			
Photovoltaics		xxxxxxxxxxxxxxxxxxxxxxxxxxxxxxxxxx				
Geothermal Aquifiers		xxxxxxxxxxxxxxxxxxxxxxxxxxxxxxxxxxxxxxxxx				
Agricultural and Forestry Wastes		xxxxxxxxxxxxxxxxxxxxxxxx				
Off-shore Wind Energy		xxxxxxxxxxxxxxxxxxxxxxxxxxxxxxx				
Advanced Conversion		xxxxxxxxxxxxxxxxxxxxxxxxxxxxxxx				
Energy Crops		xxxxxxxxxxxxxxxxxxxxxxxxxxxxxxx				
Tidal Power		xxxxxxxxxxxxxxxxxxxxx				
Wave Energy		xxxxxxxxxxxxxxxxxxxxxxxxxxxxx				
Advanced Fuel Cells		xxxxxxxxxxxxxxxxxxxxxxxxxxxxx				
Photoconversion		xxxxxxxxx				
Geothermal Hot Dry Rock		xxxxxxxxx				
Thermal Solar	xxxxxxxx					

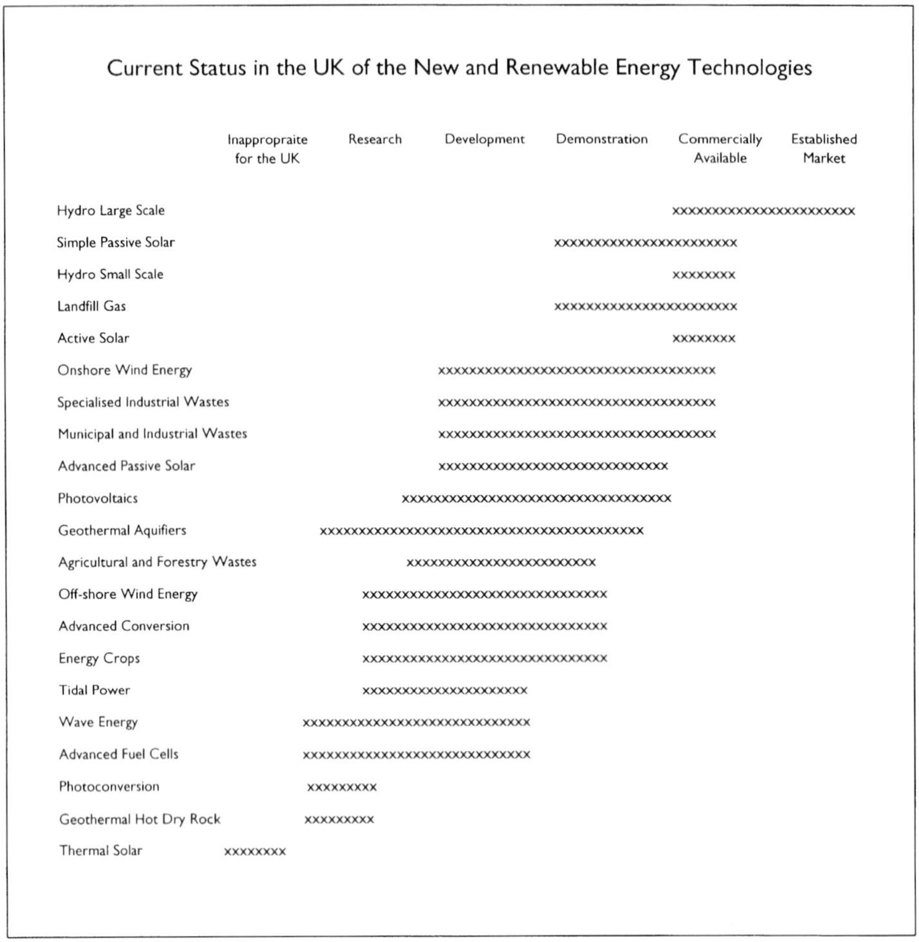

Source: Energy Paper No 62, p 6. HMSO

The cost criterion was taken as 10 pence per kilowatt-hour (p/kWh), much higher than the average current pool price of less than 3 p/kWh, but of course a major consideration in the programme is to reduce the costs of these technologies.

Estimates of the theoretical resource for the various technologies have been made and then further refined by producing a more realistic measure called the Maximum Practicable Resource. This allows for all probable constraints on using the mature technology. For example, for wind power this allows first for National Parks, housing, roads and lakes; then allowance is made for further constraints such as regulatory, sociological and environmental. Cumulative supply curves were then produced for the years 2005 and 2025 at both 8% and 15% discount rates.

Supply Curves for Electricity Producing Renewable Energy Technologies: Maximum Practicable Resource, 2005, 8% discount rate

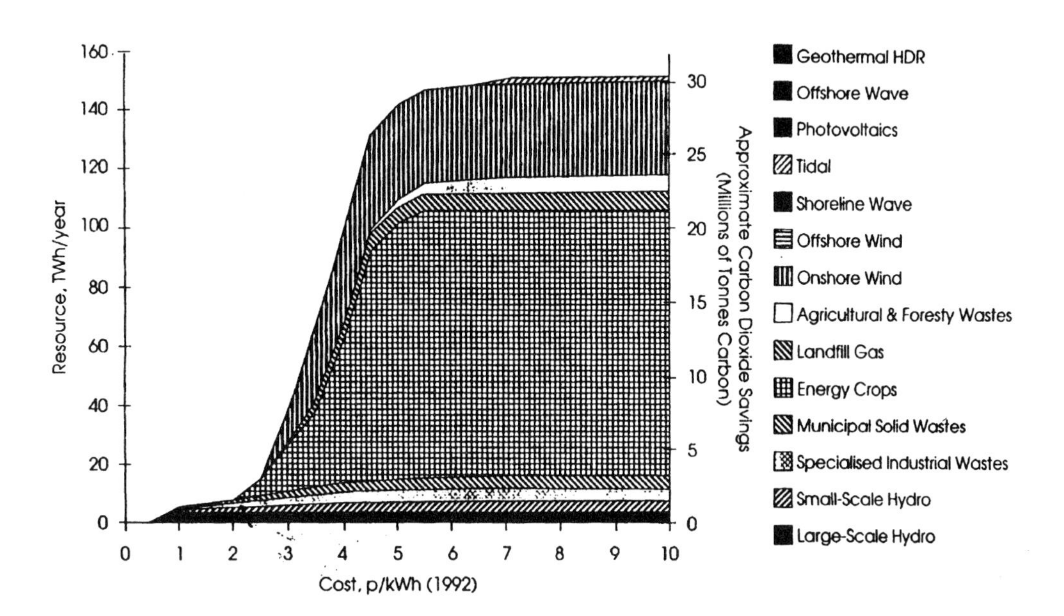

Source: Energy Paper Number 62. HMSO

The UK Government intends to collaborate internationally, especially with the EU programmes. It has then classified technologies under three heads, namely ; —

- market enablement;
- assessment, research, development, demonstration and dissemination (RDD&D);
- watching brief, which appears to mean no activity.

The policy document then gives detailed descriptions and assessments for the various types of renewables.

In October 1993 the government stated that it intended to make Orders under the NFFO arrangements for five technology bands. They were wind, hydro, landfill gas, municipal and industrial waste, energy crops and agricultural and forestry wastes for overall capacities up to about 400 MW. Financial support for renewables from the NFFO was announced[22] as being £30 million by 1993 and likely to build up to £150 million before reducing. An alternative source of official information[24] gave figures of £29 million for 1992/93 and £68 million for 1993/94. The budget for the research and allied programme known as RDD&D is given as £19.78 million for 1994/95. and expected to reduce over a ten year period. The government is expecting capital investment by industry to exceed £3000 million — with some uncertainty in the prediction — probably up to the year 2010.

So renewables could add greatly, worldwide and in the UK, to our capacity for meeting future increased energy needs. But could more be done to reduce those needs without interfering with economic development?

References

1 Quoted in *Chemistry in Britain* (Jul 1994) p.525

2 *Renewable Energy Advisory Group: Report to the President of the Board of Trade. November 1992* London: HMSO, 1992 Energy Paper no. 60

3 *Renewable energy resources. Opportunities and constraints 1990 - 2020. Report 1993* London: World Energy Council, 1993

4 *BP Statistical Review of World Energy* London: British Petroleum Co plc., 1994

5 *The World Bank and the environment. Fiscal 1993* Washington, DC, USA: The World Bank, 1993 p.91 et seq.

6 I Berkovitch 'Solar energy. Can S E Asia tap the energy of the tropical seas?' *Nusantara* (Jun 1985) pp.39,40

7 I Berkovitch 'Drying by the elements' *African Farming* (May/Jun 1985) p.33

8 'The world's first practical tidal current turbine' *IT Power News*, no. 10 (Feb 1993) and private communication

9 K Voss, W Stahl, A Goetzberger 'The self-sufficient solar house Freiburg: A building supplies itself with energy' paper in: *Solar energy. Proceedings of UK-ISES 20th Anniversary Conference* London: Solar Energy Society, 1994

10 Kevin J Brown 'The Future for Renewables' paper delivered at 'Technology in the Third Millennium: Energy' discussion meeting of The Royal Society and Royal Academy of Engineering, London, March 1994. Published in D Rooke, I Fells, J Horlock *Energy for the future* London: E and FN Spon for the Royal Society, 1995 Technology In The Third Millennium no. 6 ISBN 0419200509

11 Wolfgang Palz 'Power for the world: a global photovoltaic action plan' paper in: *Solar energy. Proceedings of UK-ISES 20th Anniversary Conference* London: Solar Energy Society, 1994

12 For example Wim Hulscher and Peter Fraenkel (introduced by) *The power guide: an international catalogue of small-scale energy equipment* London: Intermediate Technology Publications in association with University of Twente, the Netherlands, 1994 ISBN 1853391921

13. Report of World Commission on Environment and Development, chaired by Norwegian Prime Minister, Mrs Gro Harlem Brundtland. 1987. Quoted in Ref 2 above

14 'Methane - gas of the month' *Greenhouse Issues* (published London: IEA Coal Research) no. 2 (Feb 1992)

15 'Three Gorges dam project' in News Digest section *Keesing's Record of World Events* (Apr 1992) p.38862

16 Thomas B Johansson (ed.) et al. *Renewable energy - sources for fuels and electricity* Island Press, 1993 UN Report

17 *Renewable energy resources: opportunities and constraints 1990-2020. Report 1993.* London: World Energy Council, 1993

18 A R Trenka *Ocean thermal energy conversion (OTEC): a status report on the challenges* Washington DC, USA: American Chemical Society, 1993 Paper ENVR 22

19 H Tabor 'Solar ponds. Selected papers on solar radiation and solar thermal systems' in D E Osborn (ed.) *Selected papers on solar radiation and solar thermal systems* Bellingham, WA, USA: SPIE-The International Society for Optical Engineering, 1993

20 'Council decision of 13 September 1993 concerning the promotion of renewable energy sources in the Community (Altener programme)' (93/500/EEC) *Official Journal of the European Communities* (18 Sept 1993) pp.L235/41, /42, /43, /44

21 *Energy Select Committee: fourth report on renewable energy. Vol 1 March 1992* London: HMSO, 1992

b) *National Audit Office report on the Renewable Energy Research, Development and Demonstration Programme. Jan 1994* London: HMSO, 1994

c) *An assessment of renewable energy for the UK* London: HMSO, 1994 ETSU-R82

22 *New and renewable energy: future prospects in the UK. March 1994* London: HMSO, 1994 Energy Paper no. 62

23 Kevin J Brown 'Renewable energy developments in the UK during 1993' in: *British Annual Energy Review of 1993/94* London: British Energy Association, 1994

24 Personal communication from Department of Trade and Industry

25 Alison Taylor *Attitudes towards wind power* London: DTI/ETSU W/13/00354/038/REP. Also summarised in *RE View* (London, DTI) issue 22 (May 1994) pp.12,13

26 *Wind energy* London: House of Commons Welsh Affairs Committee/HMSO, 1994

27 David Hart *The Volta River Project: a case study in politics and technology* Edinburgh: Edinburgh University, 1980

Chapter 7

Conservation

Contents

- Where Can We Save?
  - Manufacturing Industry
    - Iron and Steel
    - Chemicals
    - Pulp and Paper
    - Cement
    - Agriculture
    - Residential and Commercial
    - Transport
    - Energy Conversion and Supply
    - Motor Drive Systems
- What are the Barriers to Energy Efficiency?
- Some Trends and Areas of Research
- UK Trends
- References

Conservation

"We should be good guests on earth, neither too demanding nor disturbing its delicate balance. We should allow it to renew itself for those who are to follow".

Mrs Indira Gandhi, Prime Minister of India, welcoming the World Energy Council, 1983.

Between any source of energy and its end-use there are stages of conversion — sometimes many stages. For instance we switch on a light, TV or machine, using electricity, which has been distributed along wires, from sub-stations, after generation at a power station, using in a heat cycle fossil or nuclear fuel, which had to be transported, after first being extracted from the earth. At each stage there are energy losses. Attempts to estimate the overall efficiency in using energy have led to quite startling results.

There are many critics of our wasteful styles of using energy. Yet it comes as a shock to learn that after considering the whole train of operations from energy source to end-use one researcher concluded[1] that the end-use efficiency in the USA was 2.5% To avoid any possibility of misunderstanding he added that this meant that the same final services such as providing heat, light, transport, cooking, cooling, entertainment and others, could have been obtained by using only one fortieth of the amount of energy actually used. Western Europe and Japan assessed in the same way proved to be significantly more efficient with values in the range 4 to 5%. For the world as a whole, the estimate was 3% to 3.5%. The final conclusion was that there is no fundamental technical reason why end-use efficiency could not be increased several-fold, perhaps by a factor of three, during the next century.

Early in the first chapter it was noted that energy intensity has continually greatly improved in the industrial countries. This index is the ratio of energy used to Gross Domestic Product at constant prices. It is an overall measure of changes in the efficiency of using energy in a country or an area. Broadly, this index is lowered by technological progress which generally increases efficiency. Yet, like other composite indices, the concept of energy intensity has to be applied with some understanding. If a developing country introduces heavy industry that is energy-intensive this will apparently make the energy intensity worse, though the country may be making considerable technological progress. For instance, in Nigeria commercial energy intensity is still increasing (and commercial energy is replacing traditional forms of energy source)[2] though the country is rapidly industrialising. So the pattern of change of energy intensity with time for developing countries may well be one of early increase followed by decrease but it is expected that improvement rates in this index in the developing countries will be faster than those in the industrialised countries because of technology transfer.

Though the advantages of improving energy efficiency may seem manifest, it is worth setting out explicitly reasons for actively pursuing such policies. We may start from the hard-headed or cynical view that the consumer will not pay to save energy for altruistic reasons[6] so that any increase in using high-efficiency equipment must be driven by economic or regulatory pressure. A more complete view would include all of these factors:—

- improving economic efficiency;
- reducing damaging impact on the environment;
- for many countries, reducing dependence on imported energy;
- complementing the previous point, increasing the efficiency of using scarce domestic resources of energy;
- worldwide, conserving global energy reserves, which must be finite.[3]

Final action by individual organisations will call for legal and technical assessments as well as commercial judgments but for a country as a whole, a government may intervene using various forms of pressure and guidance to promote action for improving energy efficiency.

Where Can We Save?

Studies carried out for the World Energy Council (WEC) indicated extensive potential for improving efficiency in using energy. The final report[3] picked out from the 'huge number of possibilities' a selection of examples in these important areas.

Manufacturing Industry

This sector in the advanced industrial countries is already considered to be the most efficient energy user; yet the studies showed that it had potential for improvement. Manufacturing industry uses energy for process heat and in the form of electricity. Big users are the steel, petroleum, chemicals, and paper and pulp industries. Despite large falls in energy input per unit of output notably after the oil price crisis in 1974, overall, manufacturing industry was assessed as having the potential for a further one third reduction in energy use per unit of output.

Iron and Steel

Though modern plants using the basic oxygen furnace (BOF) are integrated taking iron ore through to finished steel and have shown improvements in energy efficiency, the WEC investigators saw several possibilities for further gains. In making coke, cooling the hot coke by non-oxidising gas instead of quenching with water gives a sharp reduction in energy demand. Other technological developments are proposed but the most striking is direct steel making from ore, offering the possibility of reducing energy use by 40%.

Chemicals

This group of industries is also considered to be among the more energy-efficient and to have made big advances, but to have scope for more. Specific options proposed included these:—

- using biotechnology to speed reactions and reduce the temperatures and pressures needed;
- using catalysts to raise yields, shorten reaction times, reduce temperatures and pressures;
- improving process control by better sensors and instruments;
- improving the use of waste heat.

Systems of process integration are being introduced more widely aiming at design of processes linked with each other so that waste heat (carried by hot discharged products or intermediates) is recovered to heat up feed materials entering other processes. An important version of this approach known as Pinch Technology[4] is being extended from single processes to application to total sites. Factories involving several processes are serviced by and linked through a central utility system. The aim is then to minimise the use of fuel, to promote co-generation of electricity and process heat, and to minimise emissions and cooling. Where this approach has been applied to existing plant operating separately (retro-fitted) there are examples reported of large savings with short pay-back periods such as six months[5]. If the plant is designed from the start according to these principles higher efficiencies for lower investment costs can be achieved.

Detailed proposals were also made in respect of pulp and paper. One of the more interesting was the use of advanced pulping processes to include bio-pulping by applying enzymes derived from wood rot fungi, as well as processes using chemical pulping with fermentation.

In examining the potential for saving in cement production — where energy may account for up to half the production costs — the WEC report also included in effect a warning that some features may demand more energy. Improved environmental protection may consume energy; similarly finer grinding for stronger cements may also require greater use of electricity.

Agriculture

Energy is extensively used in agriculture both directly — in tractors, irrigation, crop drying, refrigeration, transport of crops — and in manufacturing supplies such as farm equipment and agro-chemicals. There are proposals for improvements in both these areas. One example is in irrigation where there can be gains in efficiency by using drip irrigation. This uses sensors to monitor the actual water need of plants, and computer scheduled irrigation.

Residential and Commercial

Governments and commercial interests have put out extensive publicity for methods of improving the use of energy, particularly of heat, in buildings in the developed countries in cooler climates by better insulation and instrumentation, more efficient lights and equipment. The governments have offered information, applied pressures by regulations and sometimes financial incentives.

Developing countries are often in hot climates and the WEC sees very large scope for increased use of energy in those areas for cooling, ventilation and dehumidification. This suggests an urgent need for attention to both efficient equipment and to simple means of reducing needs for ventilation and cooling. Heat gains into buildings from outside can be reduced by shade trees, awnings, shades, tinted windows, insulated windows, light coloured roofs, wall and roof insulation (housing in developing countries is often of low quality and may be constructed from uninsulated concrete blocks), and traditional heat shielding materials such as thatch.

Cooking in developing countries may be the most important use of energy and in those countries is still often supplied by burning the traditional fuels of wood, crop residues and dung. Where incomes rise people change to using modern stoves and fuels such as kerosene, LPG and electricity. Efficiency and performance increase but so does the cost. Organisations promoting technology transfer such as the voluntary group Intermediate Technology have worked with local people[17] to design simple and cheap cooking stoves of higher efficiency.

Energy intensity in cooking has declined in the developed countries. This is attributed to the combined effects of improved efficiency of equipment, switching to electric stoves and microwave ovens, and changes in the number of meals taken in the home. Microwave heating is said to be up to 10 times more efficient than the average electric oven.

For lighting, the efficiency overall is about 3%, but 10% for fluorescent lights. Though people are demanding ever better lighting, the use of electricity could be greatly reduced by speeding the change from incandescent lights to compact fluorescents — and of course by the simple measure of switching off when lights are not needed.

From time to time there are campaigns in the developed countries for improving efficiencies of the wide range of domestic electrical equipment such as refrigerators and washing machines; testers point to the great variation in efficiency between different models giving the same service indicating the scope for improved technical design. A relevant EU scheme is discussed below.

Transport

Vehicles have steadily improved in their fuel economy particularly through improved aerodynamic design reducing the drag. But there are still immense gaps between current levels of fuel consumption and the extraordinary figures of many thousands of kilometres per gallon achieved in the 'marathons' with specially constructed vehicles. These vehicles are of course far removed from designs thst could be of normal practical use yet they do suggest that further big gains are possible in fuel economy in cars in normal use.

Major improvements in the public transport system are often suggested as a means of achieving both better use of fuel and environmental objectives in many circumstances. The WEC quotes the results in Curitiba in Brazil as an example of the possibilities of an efficient bus network and land-use planning. 70% of Curitiba's population now use the system although it has a very high level of car ownership; the result is one of the lowest levels of fuel consumption per vehicle in Brazil.

Tele-conference systems and greater use of communication systems have been urged as a means of reducing the need for travelling connected with business and employment, with consequent economy in energy usage. These approaches are increasingly being adopted but the consequences in terms of travel are not yet very clear. Assessing the effects of what they call 'tele/home working' on transportation needs forms part of a new research programme[11] on the sustainable city.

News reports from developing countries with their shots of lorries belching out clouds of smoke amply confirm the expert views that there is great scope for improving their efficiency of using energy. However, much more is needed than better vehicles alone; allied needs are modern road systems, better fuel and effective repair and maintenance services.

Energy Conversion and Supply

In this area there has been what WEC calls 'fruitful experience' in plant that combines production of heat and electric power (CHP) — briefly discussed in Chapter 2 — and sometimes combines this further with district heating schemes (DH). At its best, when heat and electrical demands largely coincide with each other, CHP is claimed to be up to 90% efficient but it may still be used even where there is not an optimum balance between heat and power requirements on a site. A CHP unit can be sized to meet the base load requirements. Additional boilers can be provided that cut in and out automatically to meet extra heating demands.

There are also other methods of using heat that was previously discharged and wasted. These systems may be called co-generation or total energy; none of the terms is tightly defined and they all describe systems linking together more than one device for using heat and minimising the final discharge of heat to waste.

In recent years there has been increasing use of gas turbines for power generation, instead of the coal-fired steam driven turbines which have dominated the field. These machines have steadily increased in size and efficiency. Rolls-Royce report[7] that a new industrial machine of this type if used in simple cycle will have a thermal efficiency of 42% but the hot exhaust gases from such machines can be used to raise steam and drive a steam turbine-generator. Even with older gas turbines the combined unit — combined cycle gas turbine, CCGT — can have an overall efficiency of 55%. Adding the equipment for the steam turbine doubles the capital cost but is still cheaper than nuclear or coal-fired power stations, takes less space and is built more quickly. Power stations based on simple cycle systems can increase output by converting to combined cycle operation.

The first fuel considered is evidently natural gas. Yet these machines can alternatively burn diesel fuel, low-quality waste gases and even crude oil after some treatment. Furthermore, as we noted in the chapter on coal, coal may be gasified — there are a number of options for processes — and this gas used to fuel the gas turbine in the integrated gasification combined cycle system (IGCC). Manufacturers, sometimes collaborating with each other, are confident of raising efficiency still further.

Motor Drive Systems

A survey of pumps, fans, compressors, machine tools and other equipment driven by electric motors shows that they are often over-sized as a kind of insurance to give safety margins. Broadly, the user considers the extra capital and energy cost are acceptable to reduce the risk of failure of the equipment. The WEC investigators urged greater use of high efficiency (therefore smaller) motors, attention to reliability and voltage of power supply in developing countries, better quality of pumps, fans and the like with improved maintenance. Here too, 'smart systems' with better controls and instrumentation could often save energy.

What are the Barriers to Energy Efficiency?

Using energy more efficiently means that users could get the same service while using less inanimate energy or getting more service from the same amount of energy. Since there is so much evidence that we are generally using energy at low levels of efficiency what are the barriers to improvement?

The UK Energy Efficiency Office (EEO) analyses[8] these as barriers of information, capital and market distortion; it supplies information and selective grants to try and overcome these barriers. Both the EEO and the WEC have commented on the fact that fuels do not bear their environmental costs, sometimes called 'externalities'. These are the costs imposed elsewhere as a result of using the fuels such as damage to human health and to buildings due to pollution and acid rain, and the large contribution to atmospheric CO_2 the major man-made cause of global warming.

Professor D W Pearce has even argued[23] that the production and use of energy is inefficient because of the failure of energy prices to reflect the full costs of production and use. Full costs are defined as private costs plus external costs. The latter are the costs imposed on local, national or global communities briefly noted above. Pearce has suggested ways in which these environmental externalities would affect energy prices.

One aspect of market distortions is that the user of energy may be different from the body responsible for the energy using equipment. This applies to the builder of factories, offices or homes, or the constructor of equipment. In turn this feature emphasises the importance of government regulatory action in setting appropriate standards and checking by inspectors that they are maintained.

Governments can also, and do, intervene by setting the levels of taxation. The US is a heavy consumer of petroleum — two thirds of it for transportation — but the level of tax on gasoline is very low compared with that in West Europe which must hinder interest in improving energy efficiency beyond that required by regulations. But the Governments in the USA and the UK have both stated in the past that they do not wish to use tax policy for promoting energy efficiency, though there appears to be some change now in the UK.

Companies that are major energy users - such as steel, glass chemicals and others — naturally see the direct need for attention to energy efficiency. For other industries, the fuel costs may be only a small part of their outgoings so that this area is not seen as a very rewarding area for investing capital — which may be limited — even if the returns are satisfactory. For the same reason, directors may not wish even to give attention to improving energy efficiency and are very poorly informed.

On the energy supply side, the fuel suppliers — like any other suppliers — will naturally wish to sell more of their wares and many tariff systems include reductions for increased levels of consumption. These public utilities generally pay lip-service to promoting energy efficiency and often offer advisory services to industry and to domestic consumers, yet they will naturally benefit by supplying more of their fuel. But in the USA significant Demand Side Management (DSM) schemes have been operated[22]. These were schemes supported by the utilities to restrict energy demand. DSM aims to 'create a climate for saving energy' and includes such measures as promoting energy-efficient lighting, improved house insulation, and for industry energy audits. They were originally driven[9] by restrictions on generating capacity and local opposition to building new plant.

According to Mr Eoin Lees, Chief Executive of the UK Energy Saving Trust, the US utilities are vertically integrated, unlike the present situation in the UK, and are more strongly regulated in the US than in Britain. The Trust urges similar attention to reducing the load of UK utilities with their support and therefore argues in favour of incentives to utilities for cutting load at least equal to those for increasing their loads. The policy advocated is summarised as helping them to move from being fuel utilities to acting as energy service companies. A UK example quoted by Professor Peter F Smith[24] is that of the Merseyside and North Wales Electricity Board with its Holyhead Power Save Project. In order to avoid providing new generating capacity, this utility is reported to be subsidising low energy appliances and fittings to reduce energy consumption in local homes, commerce and industry.

Some Trends and Areas of Research

Demand for electricity is growing because of its convenience and wide range of flexibility in use but we have already commented on the generally low efficiency of conversion of energy from primary source fuels to electricity. The average efficiency of central power stations in developed economies is estimated[6] to be 35% and in the developing economies 26%. About a third of all fossil fuel is used to generate electricity. It is unfortunately expected that the growth of demand for electricity is going to increase the use of primary energy far more than any potential savings. Even the best simple cycle systems based on the steam cycle are expected to attain efficiencies of only up to 50%, and those based on gas turbines up to 58%. However, as noted above, combined cycle systems — there are now several variants under examination — and combined heat and power systems could greatly raise efficiency levels and reduce the growth in primary energy needed for generating power.

Two 'long term scenarios' have been compared and contrasted by J Masters[6].

If all world economies were to increase their consumption of electricity to the same levels as the countries of the OECD, one scenario considers the consequences with the existing mix of fuels and efficiency. Fossil fuel consumption would rise by 300% and CO_2 emissions by 480% but if the growth of consumption were met by the use of combined cycle plant of 80% efficiency the increase in fossil fuel consumption would be 180% and that in CO_2 emissions 190%.

In recent years there has been revived interest in devices called fuel cells, briefly mentioned in the chapter on coal. They are based on converting the chemical energy of fuels directly into electrical energy. So far their application has been limited by practical, technical difficulties, but they are the subject of extensive research programmes. The fuel cells currently being developed generate power by oxidising hydrogen (yielding an exhaust gas of only water vapour) and this hydrogen is obtained by processing natural gas. In a further stage of combined processing, the Coal Research section of the International Energy Agency has pointed out the possibilities of linking fuel cells for generating power with synthesis of both gas and liquid fuels based on coal.[10] If a coal gasification plant were designed on a large enough scale to meet the highest power demand on the fuel cell generating system, it would be used at other times well below its capacity. So the proposal is to consider an Integrated Gasification Fuel Cell (IGFC) yielding gas of high quality for the fuel cell, converting part of the gas stream to methanol during periods of low demand for electricity. The fuel cell can operate with methanol. Consequently, when there was low power demand, methanol would be stored for use when needed in stepping up power production at peak times. Any excess methanol could be sold.

The EU has set up a series of research, development (R & D) and demonstration programmes under a variety of acronyms — JOULE, THERMIE, ALTENER,

SAVE — to promote improved energy efficiency and the use of renewables. A further programme adopted late in 1994 is known as the Clean and Efficient Energy Technology programme. It offers up to 50% support for R&D projects and up to 40% for demonstration projects, in both cases in non-nuclear technologies. The fields of R & D include rational use of energy in buildings, industry and transport; combustion; renewable energy especially photovoltaics, wind energy and biomass; fuel cells; advanced batteries; clean coal technologies; and hydrocarbons exploration and production.

Earlier the EU Commission had proposed a carbon/energy tax specifically to promote energy efficiency, that would be fiscally neutral. Its proceeds would be used to promote this efficiency by all means, including R&D, information and subsidy for equipment that would conserve energy and/or increase efficiency of use. A majority of members agreed but some, led by the UK, opposed this form of tax and it was rejected.

During 1994 it was stated that the European Commission was preparing a Draft Directive on Integrated Resource Planning. This would include requirements for electricity and gas companies to invest in energy conservation, such as promoting insulation of buildings, somewhat along the lines of the DSM schemes quoted above. Earlier the EU had issued a Regulation[19] on a framework for introducing ecolabels on appliances. Ecolabels are being awarded to products less harmful to the environment; energy is one of the issues in judging the products. The scheme is voluntary.

UK Trends

Although efficiency in using energy in manufacturing industry has increased 'dramatically'[9] in the OECD countries it is not considered to have increased to the extent that is economically justified and environmentally necessary. For UK industry the improvement is evident in the fall in use of industrial energy from 2720 PJ (P = peta or 10^{15}) in 1973 to 1623 PJ in 1990, when the value of industrial production in real terms was almost the same. The spur was the increased cost of energy following two oil crises, and the impetus continued even when real energy prices fell, though there were other factors too — notably changes away from such energy-intensive industries as steel to others such as electronics that are not energy-intensive.

The UK government has ratified the UN Framework Convention on Climate Change which requires developed countries to aim to return emissions of CO_2 to 1990 levels by 2000 and to provide assistance to developing countries. A UK Climate Change Programme aims to save CO_2 emissions equivalent to 10 million tonnes of carbon (MtC) or 6% against the projected emissions in 2000. After extensive consultations the government has adopted a partnership approach.[12] The Government states that it is providing the fiscal, financial and regulatory framework, while business, voluntary, consumer and environmental groups and local government play their part.

The main measures include introducing VAT on domestic fuel and power (though this was originally introduced in the Spring Budget 1993 mainly to raise additional revenue), a long-term strategy of real rises in transport fuel duties, setting up the Energy Saving Trust and 'strengthening' the programmes of advice by the Energy Efficiency Office. Targets have been set for the Government's own properties, that are expected to reduce their energy use to less than 80% of 1990 levels by the year 2000. In addition, the Government is encouraging the use of CHP; the increases in CHP are expected to save about 1 MtC by 2000. As noted in Chapter 6, there is also a target for renewables of 1500 MW of new capacity subsidised by the NFFO by the year 2000, and support for R&D plus demonstration. Renewables are expected to contribute 0.5 MtC by 2000. New Building Regulations have been introduced[21] to save energy by demanding double glazing and other measures of improved building insulation; they were due to come into force in 1995.

Surveys of CHP schemes by the Department of Energy (now forming part of the Department of Trade and Industry) have shown[16] a decline in total capacity between 1977 and 1988 but a recovery since then. However, comparisons by the European Commission indicate that in 1992 the UK was one of the smaller users of CHP schemes, generating about 3% of its electricity in this way. The largest users were Denmark and the Netherlands where CHP contributed 29% of the electricity generated.

Total electricity generated by CHP in the UK appeared to have recovered by 1991 to a little above the level of that in 1977 but it is difficult to judge this because the surveys before 1991 excluded all plant owned by the public supply system, CHP/community heating schemes, CHP plants in buildings and some others. These figures are admitted in the official account not to be directly comparable. However, a further survey[16] in 1993 did show a further improvement on a comparable basis. Between 1991 and 1993 total CHP electrical capacity increased by 25% and CHP electricity production by 20%.

The Energy Saving Trust mentioned above was established in 1992 by the Government with the Regional Electricity Companies and British Gas as a non-profit making company. By both advising others and by its own projects, it has a target of promoting the saving of enough energy to reduce carbon emissions by 2.5 MtC by 2000. It was being financed by levies on gas and electricity but during 1994 there was some doubt about this source[13] of funding after objections by the gas regulator Ofgas, and the Trust was also looking for other sources.

In addition the Energy Efficiency Office which is a part of the Department of the Environment disseminates information in a Best Practice Programme and provides grants in two targeted schemes. They are:—

- The Energy Management Assistance Scheme assisting small companies to obtain consultancy advice; and
- The Home Energy Efficiency Scheme with grants to the elderly, disabled and those on low incomes to cover draughtproofing, and loft and tank insulation with restrictions on eligibility.

On the research side, three Research Councils are jointly operating a clean technology programme[20] that includes work on fuel cells and on clean combustion. For fuel cells, the major areas include applying materials science for achieving long-term reliable operation of high temperature ceramics in the cells, polymer chemistry for high performance proton conductors, and systems engineering. In the clean combustion work fundamental research is in progress on combustion processes and fluid dynamics to minimise emissions and avoid the need for clean-up.

However, some aspects of Government policy have seemed strange. The Government in its 1994 statement on global climate change[18] drew attention to transport accounting for 24% of UK CO_2 emissions and being the fastest growing source so that it was essential to reduce these emissions.

They aimed, they said, to increase the attractiveness of other modes of transport than cars, such as rail. In fact they have reduced rail subsidies, forcing fares to rise, and have broken up the integrated rail network of the single organisation British Rail, into smaller separate organisations for privatisation, where it could be more difficult to book through tickets. These measures are likely to discourage rail travel.

Critics have welcomed the progress, for example in tighter Building Regulations, but in effect commented that the new measures are too little and too late[14]. The new Building Regulations will bring new buildings in Britain to the same energy efficiency standards as those of Scandinavia in the 1960s and result in very small savings compared with the effect of current insulation standards actually being operated by builders. Surprisingly, the Government also twice blocked an Energy Conservation Bill[15] drafted with the help of the Association for the Conservation of Energy and other organisations then presented to Parliament by the Liberal Democrats with all-party support. This Bill made it a duty of local authorities to conduct energy audits and draw up energy conservation plans and priorities, with government contributions to the costs involved.

What then — internationally and nationally — are the policies being advocated to bring under control the rapidly increasing demands for energy with their potentially destructive effects?

References

1 R U Ayres 'The energy policy debate. A case of conflicting paradigms' *WEC Journal* (Jul 1992) pp.29-45

2 Nebojsa Nakicenovic *Decarbonization: doing more with less* Laxenburg, Austria: International Institute for Applied Systems Analysis, Dec 1993 Working Paper WP-93-076

3 *Energy for tomorrow's world* London: Kogan Page for the World Energy Council, 1993

4 V R Dhole, B Linhoff 'Total site targets for fuel, co-generation, emissions, and cooling' paper to conference 'ESCAPE-2' held 5-7 October 1992 Toulouse, France, 24th European symposium of the Working Party on Computer Aided Process Engineering published in: *Computers and Chemical Engineering* (United Kingdom) vol. 17: supp. 1993 pp.S101-S109

5 S Dincer, I Sentarh 'Heat-integration retrofit study of a petroleum refinery' *Applied Energy* 38(4) 1991 pp.253-262

6 J Masters 'Oil and gas utilisation: limits of efficiency and their impact on demand' paper delivered at 'Technology in the Third Millennium: Energy' discussion meeting of The Royal Society and Royal Academy of Engineering, London, March 1994. Published in D Rooke, I Fells, J Horlock *Energy for the future* London: E and FN Spon for the Royal Society, 1995 Technology In The Third Millennium no. 6 ISBN 0419200509

7 Ron Haywood 'New technologies in power generation' *Rolls-Royce Magazine* no. 62 (Sept 1994) pp.11-14

8 J Hobson *The barriers to improvement* paper to Annual Energy Forum 1994. Due to be published by the British Energy Association

9 E W Lees 'Energy efficiency and approaches to it' paper delivered at 'Technology in the Third Millennium: Energy' discussion meeting of The Royal Society and Royal Academy of Engineering, London, March 1994. Published in D Rooke, I Fells, J Horlock *Energy for the future* London: E and FN Spon for the Royal Society, 1995 Technology In The Third Millennium no. 6 ISBN 0419200509

10 David H Scott *Advanced power generation from fuel cells — implications for coal* London: IEA Coal Research, July 1993 IEA CR/59

11 *Effects of tele/home working on transportation needs.* Research topic within 1995 programme 'Towards the Sustainable City' of the Engineering and Physical Sciences Research Council (EPSRC, North Star Ave, Swindon, SN2 1ET, UK)

12 Linda Makin 'The Government's response to climate change.' *RE View* issue 22 (May 1994)

13 Information brochures and private communication from Energy Saving Trust, 11-12 Buckingham Gate, London SW1E 6LB

14 Statement by Andrew Warren, Director, Association for the Conservation of Energy, 9 Sherlock Mews, London W1M 3RH

15 *Energy Conservation A Bill to make provision for the conducting of energy audits and the drawing up of energy conservation plans...*, presented by Mr A J Beith, London: HMSO, 1993 Bill 11, 139050, 51/2

16 'Combined heat and power' in: *Digest of United Kingdom Energy Statistics 1993 Annex C.* London: HMSO. 1993. and 'Combined heat and power' in: *Digest of United Kingdom Energy Statistics 1994. Annex C.* London: HMSO, 1994

17 Intermediate Technology Development Group, 103 Southampton Row, London WC1B 4HH

18 *Global Climate Change* 3rd ed., London: Department of the Environment in association with the Meteorological Office/HMSO, 1994

19 'Council Regulation (EEC) No 880/92 of 23 March 1992 on a Community Eco-Label Award Scheme' *Official Journal of the European Communities* (11 Apr 1992) pp.L 99/1-7

20 *Directory of Clean Technology Research.* Swindon: Clean Technology Unit, 1994 (Biotechnology and Biological Research Council, Engineering and Physical Sciences Research Council, and the Economic and Social Research Council - North Star Avenue, Swindon, Wilts, SN2 1ET)

21 *Building Regulations (Amendment) Regulations 1994* London: HMSO, 1994 SI No 1850 ISBN 0110448502

22 For instance the conference International Energy Conference on Demand-side Management; A Current and Future Resource. Paris, France. Held in 1991 in Copenhagen, Denmark. OECD/IEA, 1992

23 D Pearce 'Costing the environmental damage from energy production' paper delivered at 'Technology in the Third Millennium: Energy' discussion meeting of The Royal Society and Royal Academy of Engineering, London, March 1994. Published in D Rooke, I Fells, J Horlock *Energy for the future* London: E and FN Spon for the Royal Society, 1995 Technology In The Third Millennium no. 6 ISBN 0419200509

24 Peter Smith 'Out in the cold indoors' *Guardian* (24 Jun 1994)

intentionally left blank

Chapter 8

Policies

Contents

- The Four Cases [of the WEC study]
  - Case A High Growth
  - Case B Reference
  - Case B1 Modified Reference
  - Case C Ecologically Driven
- Actions Proposed
- Japanese 'Proposals for a New Earth'
- EU Energy Policy
  - The Internal Market
  - External Relations
  - Minimising Impact on the Environment
- UK Energy Policies
  - The Government
  - Labour Party Policy
  - Other Proposals: Liberal Democrats
  - Academic and Other Views
- And a Final Word
- References

Policies

"It was not enough for man to love his neighbour; he must also learn to love his world, which he kept destroying....it was essential to conclude a political and ethical contract with nature".

Boutros Boutros-Ghali, Secretary-General of the United Nations. Concluding speech at the Rio Earth Summit, 1992.

153 nations signed the Framework Convention on Climate Change at the Rio Earth Summit in 1992.[1] This Convention, it may be recalled included important commitments. All nations agreed to operate national programmes to limit emitting of greenhouse gases and to protect carbon sinks such as forests. But the developed countries also accepted a commitment to give a lead. They would return emissions of CO_2 and other greenhouse gases to 1990 levels by the year 2000. They would also provide new and additional resources to help developing countries to meet their commitments. This pledge included agreeing to transfer new and energy-efficient technologies — an important feature since emissions of CO_2 and other greenhouse gases from developing countries are expected to exceed those from the present industrial countries early in the next century.

How did the energy experts assess the possibilities of meeting these targets? Peering into the metaphorical crystal ball, the World Energy Council (WEC) had earlier begun studies of four main future possibilities. Following on decades of previous studies, the Council set up a Commission in 1989 'for a more detailed examination of the parameters relevant to future energy developments'. This Commission consulted energy practitioners, environmentalists and economists worldwide and developed four Cases of possible future energy demand[2] taking them up to the year 2020 (with a more tentative look at possibilities to 2100). Its aim was stated to be identifying a realistic framework for supplying adequate sustainable energy at acceptable costs, with socially acceptable care and protection of the environment.

The Four Cases

Different assumptions underlie the four Cases in respect of economic development, improvements in energy efficiency, speed of disseminating technology from the industrial to the developing countries, and in resolving institutional issues. They are taken to illustrate future possibilities, and are not predictions. What will actually happen naturally depends on what consumers and policy-makers choose to do. For the rate of growth of population — as we saw earlier a key factor — the Commission took the median figure of the Base Case of the UN and World Bank. This estimates an additional 2.8 billion (thousand million) people between 1990 and 2020 (5.3 billion to 8.1 billion). Another important assumption is that energy intensity — broadly the increase in energy demand per one per cent increase in GDP — reduces faster in future than ever in the past.

Case A

High Growth. This assumes an average world economic growth rate of 3.8% pa. Within this average, the growth in developing countries is more than double that of the industrial countries. Overall, both energy demand and annual CO_2 emissions in 2020 work out to be almost double the figures for 1990.

Case B

Reference. Here, based on recent experience, the WEC has modified and updated a proposal made in 1989 for a Moderate Economic Growth Scenario. It postulates world economic growth at 3.3% pa and global energy demand rising over 50% over the period of the study. CO_2 emissions rise by 44%.

Case B1

Modified Reference. Since it was thought possible that energy intensity might be reduced later and more slowly in the developing countries and the former Soviet bloc than had been assumed for the other Cases, this modified reference Case was introduced as a variant on Case B. Economic growth rates were taken to be the same as in the reference Case but the effects of the assumed slower reduction in energy intensity led, as would be expected, to higher calculated energy demand and higher CO_2 emissions.

Case C

Ecologically Driven. Suppose we all took seriously ecological issues and the dangers of rising CO_2 in the atmosphere. Suppose therefore that there was a massive drive to raise energy efficiency including faster use of new renewables and gas, while we speeded up technology transfer between industrial and developing countries. Case C — understandably described as 'politically challenging' — is based on these suppositions; they correspond with a rate of reducing energy intensity far higher than anything hitherto achieved. Yet, taking into account increased population and rising expectations in the developing world, even these almost idealistic conditions result in energy demand rising 28% over the relevant period. However, the presumed switch to new renewables, natural gas and an increase in nuclear, give a calculated increase of only 6% in the emissions of CO_2 by the year 2020.

Main Characteristics of the Four WEC Energy Cases

Case	A	B_1	B	C
Name	High Growth	Modified Reference	Reference	Ecologically Driven
Economic	High	Moderate	Moderate	Moderate
Growth % pa				
OECD	2.4	2.4	2.4	2.4
CEE/CIS	2.4	2.4	2.4	2.4
Dcs	5.6	2.4	4.6	4.6
World	**3.8**	**3.3**	**3.3**	**3.3**
Growth Per Capita				
OECD	Moderate	Moderate	Moderate	Moderate
CEE/CIS	Moderate	Moderate	Moderate	Moderate
DCs				
Asia	Very High	High	High	High
Sub-Sahar. Africa	Moderate	Low	Low	Low
Most Others	High	Moderate	Moderate	Moderate
Energy Intensity	High	Moderate	High	Very High
Reduction % pa				
OECD	-1.8	-1.9	-1.9	-2.8
CEE/CIS	-1.7	-1.2	-2.1	-2.7
DCs	-1.3	-0.8	-1.7	-2.1
World	**-1.6**	**-1.3**	**-1.9**	**-2.4**
Technology Transfer	High	Moderate	High	Very High
Energy Efficiency Improvement				
OECD	High	High	High	Very high
CEE/CIS	Moderate	Moderate	High	Very high
DCs	Moderate	Moderate	High	Very high
Institutional Improvements (World)	High	Moderate	High	Very high
Possible Total Demand (Gtoe) (1990 = 8.7 Gtoe)	Very high 17.2	High 16.0	Moderate 13.4	Low 11.3
CO_2 Emissions from Fossil Fuel (GtC) (1990 = 5.9 GtC)	11.5	10.2	8.4	6.3

Source: Energy for Tomorrow's world, Kogan Page for the WEC, 1993

However, going beyond scenarios and Cases, did the experts of the WEC have any opinion on the most likely energy demand? The calculated figures for the four Cases ranged from 11.3 Gtoe pa in the year 2020 for the almost idealistic ecologically driven Case C to 17.2 Gtoe for the high growth Case A. Current energy policies and consumer behaviour led them to consider the more likely outcome as lying between 13 and 16 Gtoe. The results of these massive international labours have been summarised in a table.

Not all of the carbon dioxide emissions remain in the atmosphere. There are complex interchanges between the atmosphere and various 'reservoirs'[12], Carbon 'cycles' between the atmosphere, the oceans, the ocean biota (living things), the soil, and the land biota. For several thousand years the exchanges were very constant and CO_2 concentration remained at about 280 parts per million by volume (ppmv) but since the Industrial Revolution this concentration has increased to about 355 ppmv due to increased combustion of fuels and the destruction of forests. Based on researches of these carbon cycles and the changes in CO_2 concentrations, mathematical models have been developed and estimates made for probable concentrations on various assumptions up to the year 2100. One of these models, known as MAGICC, applied by the Climatic Research Unit, University of East Anglia, to the CO_2 emission figures from commercial fossil fuel resulting from the four WEC Cases gave these calculated values, adapted from Ref 1.

Year	1990		2020 Ccalculated		
Case		A	BI	B	C
CO_2 (ppmv) concentration	355	434	426	416	404

ppmv is parts per million by volume

None of the Cases actually stabilises these concentrations from fuel combustion at the 1990 level. Three of the Cases are well above and only the very optimistic Ecologically Driven Case results in global CO_2 emissions at all close to the 1990 level.

Actions Proposed

It bears repeating that these are considered future possibilities — though reasonably based — not forecasts but they present so ominous a picture that the WEC advised 'as a matter of necessary prudence' the following approach:

- Recognising the uncertainties, the need for intensified research;
- The need to raise energy efficiency where justified by cost/benefit analysis and to increase energy conservation;
- Applying rational adaptation now, since if the hypotheses are justified the world is probably past the point where global climate change can be avoided;
- To achieve the necessary speed and effectiveness governments must be involved for stimulus and leadership, using market instruments such as tradable emission permits and road user pricing.

These are overall global requirements but different regions will have different priority needs. In the poorer developing countries the first priority is access to sufficient affordable energy. Present levels of energy consumption inhibit even minimal economic growth. These countries also have many other urgent problems such as deforestation, soil erosion, uncontrolled growth of cities, industrial pollution, scarcity of water — all calling for investment. These needs are in addition to those for raising efficiency in providing and using energy, so that they have massive needs for investment.

The countries of the former Soviet bloc are seen as needing to modernise and expand their existing supply structures and to rationalise their use of energy, as well as introducing regimes that foster investment.

Industrial countries need to secure greater energy efficiency but they also have responsibilities to improve continuously technologies deployed in their own countries and elsewhere.

Longer term, the huge increase expected in population will mean accelerated consumption of fossil fuels. This will result in relying ever more on coal which has by far the largest reserves. Ultimately — probably a century on — there will be a shift to synthetic gas and to other fossil resources briefly noted earlier, such as tar sands and shale oil, with greater difficulties in terms of cost, technology and environmental impact.

The WEC emphasises the global interconnections between countries and their problems. It poses a challenge to world institutions 'to secure broad progress on all fronts rather than allow sectoral or political difference to inhibit necessary progress.'

Its recommendations are grouped under these main headings: -

- Ensuring the functioning of energy markets — liberalising markets, albeit within government regulatory frameworks;
- Long-term oriented research and development — including areas not yet perceived to be justified by economic returns, and also long-distance energy transport facilities;
- Improving environmental quality — promoting economic growth in ways that protect the environment, notably with precautions to reduce emission of greenhouse gases;
- Finding a path to sustainable energy development — placing energy issues in their broadest social and institutional context, recognising that people seek services that energy provides, not energy as such.

Japanese 'Proposals for a New Earth'

In Japan, advisory bodies to the famous Ministry of International Trade and Industry (MITI), followed up a Five Year Economic Plan adopted by the government, with their own further suggestions. Published towards the end of 1992, these proposals[9] took into account both the government plan and the international consensus reached at the Earth Summit on the concept of sustainable development. Basing themselves on what they called a 'Policy Triad' for the Environment, Economy and Energy, they put forward these 'Fourteen Proposals for a New Earth' — each of them being discussed in considerable detail. They were addressed to national and local government, to business, and to consumers:—

- Maximize energy efficiency in all sectors by supporting the introduction of energy-efficient equipment and revising the Energy Conservation Law;
- Facilitate building of district energy systems to maximize use of waste heat;
- Facilitate increased use of nuclear energy and of new and renewable energy sources;

- Give strong initial support for voluntary efforts by business and a legal framework to speed the transformation to an environment-friendly socio-economic structure. Expand recycling of resources;
- Offer information on preferable types and environmental effects of consumer behaviour; create institutions that support environmental-friendly consumer behaviour;
- Create the right social infrastructure, including systems for effective energy use and for waste recovery and recycling tailored to regional characteristics;
- Promote recycling by identifying precise bottlenecks for each recycled resource and provide detailed responses;
- To conserve energy and resources, lengthen product life, improve quality. Improve efficiency of distribution, promote R & D in distribution and in energy-efficient vehicles;
- Accelerate development of photovoltaic cells, fuel cells, but also of innovative energy technologies for the medium and long term. Develop CO_2 fixation, environment-friendly production processes, and substances with a low environmental load such as CFC substitutes, biodegradable plastics;
- Expand the Green Aid Plan encouraging self-help by developing countries by energy- and environment-related international cooperation;
- Implement the Framework Convention on Climate Change and take the initiative in forming an international network;
- To harmonize trade and environmental policy, contribute actively to the work of GATT;
- Expand technology transfer as part of 'New Earth 21' — a scheme to disseminate environment related technologies — and using such organizations as the Global Environment Facility;
- Accelerate development of international agreements by international consensus on environment problems.

EU Energy Policy

Largely following proposals by the Commission[3] on the internal energy market, the EU is developing an active energy policy. It is based on these aims: —

- Developing an internal market in energy;
- Developing external energy relations;
- Minimising the negative impact on the environment of using and producing energy.

The Internal Market

Here the EU wishes to develop open competition in electricity and gas with two important provisos. There must be security of supplies and the suppliers must meet a 'public service obligation of universal, uninterrupted provision.' In 1990/91 the Council adopted Directives on Price Transparency[4], Transit of Electricity and Transit of Gas, while the Commission later proposed guidelines[5] on the trans-European energy networks. Transparency in this context means giving industrial customers access to aggregate data on charges made by gas and electricity utilities to all classes of customers throughout the Community. The guide-lines on networks covered electrical and natural gas networks, in each case promoting inter-connection and inter-operability.

Recognising the disturbance to existing procedures, the Commission has called for a gradual approach allowing industry to adjust in an orderly way; the Community must not impose a rigid system but leave it to member States to opt for the best way to introduce the new arrangements, avoiding 'excessive regulation'. The stated purpose is to define a framework. However, coal subsidies are to be limited. A Directive was proposed to help create a competitive framework for licensing oil exploration and production.

External Relations

Countries of the European Free Trade Area (EFTA), in accordance with the European Economic Area Agreement (effective from Jan 1993) are generally adopting EU rules on energy. In dealings with Eastern Europe and the former Soviet Union, the main approach is through the European Energy Charter of December 1991; by the end of 1992 it had been signed by 47 countries. Its aim is to promote economic recovery in the East European countries by joint efforts at developing energy resources and modernising its industries. The Commission has made detailed proposals on its application[14]. A further area of study is how to ensure security of supply to the EU when oil is likely to continue to provide 45% of energy supply and over 70% of this is imported. Meeting this challenge demands a good political framework but may also mean creating a strategic oil reserve.

Minimising Impact on the Environment

As we noted in the first chapter, the EC decided in 1990 that its CO_2 emissions would be stabilised by the year 2000 at the same levels as in 1990. A range of programmes have been adopted on R & D (JOULE), promoting energy technologies (THERMIE)[17], energy efficiency (SAVE)[16], energy labelling[7], renewable energy sources (ALTENER)[15]. There have also been proposals[10] for a wide range of measures to minimise the impact of transport on the environment. They include favouring public over private transport, reducing the level of motor traffic in cities, cost-charging measures and further research and development. A Commission proposal for a carbon/energy tax[8] remained, at least up to autumn 1994, still-born largely due to UK objections. This tax was designed to be fiscally neutral. All proceeds would be used for promoting energy efficiency and shifting demand away from the most polluting fuels. Similar proposals were being considered in the other industrial countries outside the EU and the Commission intended that the tax should be introduced, in effect, simultaneously by the EU and its main competitors. To encourage the development of a biofuels industry — that is motor fuels of agricultural origin — the Commission also put forward a Directive[11] for all Member States to reduce excise duties on these fuels to not more than 10% of the rate for the corresponding fuel from petroleum. Covering ethanol, methanol, vegetable oil, and chemically modified vegetable oil, the aim was to contribute a further step to security of energy supply and to the stabilising of CO_2 emissions. Combustion of biofuels (as we noted in the preceding chapter) is carbon-neutral since their carbon has been 'fixed' in recent times and forms part of the current active carbon cycle. When the biofuels are burnt they produce no more emissions of CO_2 than they would if they rotted away naturally.

Carrying forward the plans for international co-operation on fusion research discussed in the chapter on nuclear energy, the EU late in 1994 concluded an agreement[6] with Canada, Japan and the USA.

UK Energy Policies

The Government

Aspects of UK Government energy policy have been noted in earlier chapters, so that the features given here include some degree of recapitulation. The aim of the Government's energy policy[22] is to ensure secure, diverse, sustainable supplies of energy in forms that people and businesses want, at competitive prices. In its view this will best be achieved by means of competitive energy markets working within a framework of law and regulation to protect health, safety and the environment. It explicitly rejects planning for energy production and consumption but accepts the role of monitoring the way the market works.

A section of the policy entitled 'Measures to reduce energy consumption' appears to include no such "measures" but refers to estimates by the Government's Energy Efficiency Office (EEO) of scope for improvement, and the setting up of the Energy Saving Trust. Growing energy demand for transport is acknowledged to be a major contributor to overall increases in energy demand. The only comment added is 'continuation of such growth is therefore a significant factor in considerations of a sustainable pattern of energy use.' In a separate reference to transport policy[22] the principal measure for reducing transport emissions of CO_2 is a commitment year on year to real increases in fuel duties of 5%. In considering the way forward it is acknowledged that 'further measures will be necessary by Government to influence the rate of traffic growth and provide a framework for individual choice which enables environmental objectives to be met.'

On energy prices, the introduction of Value Added Tax (VAT) is stated to be part of the action taken to reduce demand and return CO_2 emissions to 1990 levels by 2000. The EC proposal for a carbon/energy tax is rejected.

In respect of measures to reduce the environmental impact of energy supply, the policy notes the value of switching fuels from coal to gas when generating electricity, in halving of CO_2 emissions per unit of electricity, and the further gain in using combined cycles. Clean coal technologies receive qualified approval. Combined Heat and Power (CHP) is being promoted as a priority area by the EEO and the Energy Saving Trust. The target for installed capacity in 2000 is increased from 4000 MW to 5000 MW corresponding to a reduction of CO_2 emissions of about 1 million tonnes of carbon.

Nuclear power is noted as a source of energy free from the pollutants resulting in acid rain and global warming. The Government declares it is committed to maintaining the nuclear option for the future, provided it can remain economic, safe and maintains high standards of environmental protection. The nuclear industry is held responsible for dealing with radioactive waste, planning for decommissioning and securing the necessary funds.

In respect of new and renewable energy, the policy is 'to stimulate the development of new and renewable energy sources wherever they have prospects of being economically attractive and environmentally acceptable.[18]' These sources are intended to contribute to:—

- diverse, secure and sustainable energy supplies;
- reducing the emission of pollutants;
- encouraging internationally competitive industries.

The policy will take account of those factors that influence business competitiveness. The aim is to work towards 1500 Megawatts Declared Net Capacity (MW DNC) of new electricity generating capacity from renewable sources for the UK by 2000. The main component of the current contribution of renewables in the UK (1% of primary energy, about 2% of electricity supply) is 1200 MW of hydro-based capacity in Scotland.

The policy refers to the subsidy provided by the Non-Fossil Fuel Obligation (NFFO) — though it does not indicate that only 6% of this is used for renewables, the balance being for the nuclear industry. There is a supporting research and development programme and the UK participates in the European Union's international research programmes. Technologies have been assessed and classified in terms of their prospects.

- Those with the best, early prospects are judged to be onshore wind, small-scale hydro, landfill gas, waste combustion and passive solar design;
- In the medium term it is energy crops, using advanced conversion technology, that are considered to have commercial prospects;
- Active solar, decentralised photovoltaics and photoconversion are thought to have possibilities but need further assessment;
- Technologies rated as having poor commercial prospects are tidal power, offshore wave, offshore wind, centralised photovoltaics, geothermal and new

large-scale hydro. 'Very limited prospects' were attributed to shoreline wave technologies.

Potential sales overseas are seen as enormous and steadily growing, justifying development of a large renewable energy industry.

Labour Party Policy

During 1994 the Labour Party held strong leads of around 20 points in polls of opinion and voting intention, so that its policies were increasingly considered as those of a potential government. Its energy policy, forming part of a wider report of a commission on the environment[13], was stated to be sustainable and self-financing. These were its main features: —

- A national programme of energy efficiency work offering householders energy services on a no-initial cost basis. This will save millions of tonnes of carbon dioxide emissions, and create up to 50,000 new jobs; householders accepting a 'package' of energy efficiency measures will pay more for energy supply but on terms that will leave them better off because of the reduced energy use.
- Higher insulation standards within the building regulations;
- Aiming to generate 10% of electricity from renewable sources by 2010 and increase the use of combined heat and power stations;
- Support for clean coal technology;
- Operating a 'presumption' against approval for new opencast coal mining;
- Not building any new nuclear power stations.

The energy efficiency programme has the linked objectives of combating fuel poverty, reducing carbon dioxide emissions and generating substantial long-term employment. In the section on global sustainable development, there is a specific objective of reducing UK CO_2 emissions by 20% by 2010. The report of the Renewable Energy Advisory Group (REAG) is quoted with broad approval and the Labour policy accepts the idea of a specific renewables obligation, increasing its proportion of the current NFFO, and increasing the energy R & D effort into renewables. In addition, overseas aid should be targeted to assist recipients to protect their environment.

Other Proposals: Liberal Democrats

An important part of Liberal Democrat policy has been the Energy Conservation Bill briefly discussed in the preceding chapter. This had all-Party support including originally that of the Conservative chairman of the Commons Environment Committee (Robert Jones M P). However, it was opposed by the government and twice 'talked out'. Later Robert Jones was appointed Minister for Energy Conservation and changed his views to support of the government position — that the proposals would place an unnecessary burden on local authorities. The (rejected) Bill made it the duty of local authorities, as energy conservation authorities, to survey, carry out consultations then draw up energy conservation plans and priorities for residential properties in their areas. The Secretary of State would publish a timetable for implementing the plans and make contributions to their costs.

Other 'key policies' include these points: —

- More action to counter and research into the greenhouse effect, including fuel switching from coal to gas, taxing carbon dioxide emissions and protecting the world's forests;
- establishment by the government of the appropriate framework within which the market operates — such as taxing polluting activities, using grants — to conserve finite resources of primary fuels;
- a target of 30% reduction in carbon dioxide emissions from the UK by the year 2005, with lower emissions of other greenhouse gases;
- withdrawing all nuclear power stations from service by the year 2020 at the latest, and earlier if feasible.

Academic and Other Views

Two academics dealing with energy policy and energy conversion — Professors N Lucas and Ian Fells — have both criticised reliance on the market as a basis for energy policy and themselves proposed[19] to the Parliamentary Group for Energy Studies a rational UK energy policy. The mix of fuels advocated for the early years of the next century to meet considerations of security of supply and needs of the environment is nuclear 25%, coal 55%, gas 15%, and renewables 5%. But responsibility for energy policy, they urge, should be restored to political institutions and a single Energy Agency created to monitor and implement all aspects of energy policy. In earlier contributions that both ably summarised the damage due to current and future use of fossil fuels, and their main causes including market failures, the 'obvious strategy' is given briefly[20] in these terms: —

- use all energy forms more efficiently;
- switch to renewable energy such as wind, tidal and 'of course' hydroelectric;
- 'more contentiously' expand nuclear power, develop efficient public transport systems, eliminate the motor car;
- 'even more difficult' — try to curb population growth.

The Commons Environment Committee at the end of 1993 criticised the Government for failing to develop a clear strategy to conserve energy and made some suggestions of its own[24]. Some of the VAT income, for instance, could be spent on energy conservation creating jobs, improving the health of pensioners and children, reducing the taxpayers' subsidy to poor households. New gasfired stations for generating electricity were wasting heat; electricity regulations should be changed to enforce introducing CHP in suitable places. Regulations for the privatised gas and electricity industries fail to encourage enough energy saving. The Committee set out a range of much stricter energy efficiency requirements for the government, including new approaches and attitudes.

Since the use of energy sources by transport is known to be the fastest growing contributor to atmospheric pollution, the relevant views of the Royal Commission on Environmental Pollution are of key importance. Ways must be found, they declared[23], to make the longer term development of transport environmentally sustainable. A summary of their objectives and targets is shown in the table.

Though they endorse the transport framework put forward by the Government, they conclude that the price of fuel needs to double by 2005. This will provide the incentive for manufacturers to develop more fuel-efficient cars and lorries. Part of the revenue could be used to fund improvements in public transport. 'The prize is a transport system which will be much less damaging to health and the environment, and at the same time more efficient in providing the access people want for work and for leisure.'

Transport and the Environment. Objectives and Targets

A: To ensure that an effective transport policy at all levels of government is integrated with land use policy and gives priority to minimising the need for transport and increasing the proportions of trips made by environmentally less damaging modes.

B: To achieve standards of air quality that will prevent damage to human health and the environment.

B1: To achieve full compliance by 2005 with World Health Organization (WHO) health-based air quality guidelines for transport-related problems.

B2: To establish in appropriate areas by 2005 local air quality standards based on the critical levels required to protect sensitive ecosystems.

C: To improve the quality of life, particularly in towns and cities, by reducing the dominance of cars lorries and providing alternative means of access.

C1. To reduce the proportion of urban journeys undertaken by car from 50% in the London area to 45% by 2000 and 35% by 2020, and from 65% in other areas to 60% by 2000 and 50% by 2020.

C2: To increase cycle use to 10% of all urban journeys by 2005, compared to 2.5% now, and seek further increases thereafter on the basis of targets to be set by the government.

C3. To reduce pedestrian deaths from 2.2 per 100,000 population to not more than 1.5 per 100,000 population by 2000, and cyclist deaths from 4.1 per 100 kilometres cycled to not more than 2 per 100 million kilometres cycled by the same date.

D: To increase the proportions of personal travel and freight transport by environmentally less damaging modes and to make the best use of existing infrastructure.

D1: To increass the proportion of passenger-kilometres carried by public transport from 12% by 2005 and 30% by 2020.

D2: To increase the proportion of tonne-kilometres carried by rail from 6.5% in 1993 to 10% by 2000 and 20% by 2010.

D3: To increase the proportion of tonne-kilometres carried by water from 25% in 1993 to 30% by 2000, and at least maintain that share thereafter.

E: To halt any loss of land to transport infrastructure in areas of conservation, cultural, scenic or amenity value unless the use of lthe land for that purpose has been shown to be the best practicable environmental option.

F: To reduce carbon dioxide emissions from transport.

F1: To reduce emmissions of carbon dioxide from surface transport in 2020 to no more than 80% of the 1990 level.

F2: To limit emissions of carbon dioxide from surface transport in 2000 to the 1990 level.

F3: To increase the average fuel efficiency of new cars sold in the UK by 40% between 1990 and 2005, that of new light goods vehicles by 20%, and that of new heavy dutty vehicles by 10%.

G: To reduce substantially the demands which transport infrastructure and the vehicle industry place on non-renewable materials.

G1: To increase the proportion by weight of scrapped vehicles which is recycled, or used for energy generation, from 77% at present to 85% by 2002 and 95% by 2015.

G2: To increase the proportion of vehicles tyres recycled, or used for energy generation, from less than a third at present to 90% by 2015.

G3: To double the proportion of recycled material used in road construction and reconstruction by 2005, and double it again by 2015.

H: To reduce noise nuisance from transport.

H1: To reduce daytime exposure to road and rail noise to not more than 65 $dBL_{Aeq\ 16}$ at the external walls to housing.

H2: To reduce night-time exposure to road and rail noise to not more than 59 $dBL_{Aeq\ 8h}$

Source: Transport and the Environment. HMSO

And a Final Word

It is a truism that all living systems are interdependent with the environment but one leading scientist has gone well beyond this in putting forward what he calls the 'Gaia hypothesis', named after Gaia the Greek Earth goddess. In a series of publications[21], J E Lovelock and collaborators have posited that 'Life, or the biosphere, regulates or maintains the climate and the atmospheric composition at an optimum for itself'. In essence, this argues with illustrations from the history of the Earth that the ecosystems on the Earth are so tightly coupled to their physical and chemical environments that the whole could be considered as virtually one organism. Feedback mechanisms, it is suggested, result in self-regulation.

John Houghton, a former Director of the Meterorological Office, quotes this in order to issue a warning[12] against using this hypothesis as an excuse for taking no action on the risks of global warming. The hypothesis itself is under debate in the scientific world but even if it were accepted in full, it would still not mean that we could ignore the vulnerability of the environment to major disturbances and the risks to human life.

Lovelock himself has warned that Gaia is 'no doting mother, tolerant of misdemeanours....Her unconscious goal is a planet fit for life. If humans stand in the way of this we shall be eliminated..'. And Houghton adds his own caveat, much along the lines of the quotations from Indira Gandhi and Boutros Boutros-Ghali at the heads of the last two chapters. Humans have already severely damaged the environment, he notes, with a call for an attitude of stewardship, on both practical and moral grounds. Nature is vulnerable to the large changes that human activities are now able to generate. We are endangering ourselves and future generations.

If this is remotely near true we need urgently to develop energy policies that truly bring us nearer to living in conformity with the needs of the rest of nature.

References

1 Report of the Intergovernmental Negotiating Committee for a Framework Convention on Climate Change on the Work of the Second Part of its Fifth Session, held at New York from 30 April to 9 May 1992. Quoted in *Sustainable Development. The UK Strategy* London: HMSO, Jan 1994 Cmnd. 2426

2 *Energy for tomorrow's world* London: Kogan Page for the World Energy Council, 1993

3 *The Internal Energy Market* Brussels: Commission of the European Communities, May 1988 COM (88) 238. And a progress report: Brussels: Commission of the European Communities, May 1990 COM (90) 124

4 'Council Directive of 29 June 1990 concerning a Community procedure to improve the transparency of gas and electricity prices charged to industrial end user' Brussels: Commission of the European Communities, July 1990 *Official Journal of the European Communities* p.L185/16

5 'Proposal for a European Parliament and Council Decision laying down a series of guidelines on trans-European energy networks (94/C 72/04)' COM (93) 685 final - 94/0009 (COD) *Official Journal of the European Communities* no. C 72/10 (Mar 1994)

6 *Conclusion by the Commission on behalf of the European Atomic Energy Community pursuant to Article 101 paragraph 3 of the Treaty establishing the EAEC of the International Energy Agency implementing agreement on a cooperation programme on nuclear technology of fusion reactors with the government of Canada, the Japan Atomic Energy Research Institute (designated by the government of Japan) and the government of the USA* Brussels: Commission of the European Communities, 8 September 1994 Sec (94) 1476 Final

7 'Council Directive on the Indication by Labelling and Standard Product Information of the Consumption of Energy and Other Resources by Household Appliances' Brussels: Commission of the European Communities, October 1992 *Official Journal of the European Communities* p.L 297

8 *Proposal for a Council Directive introducing a tax on Carbon Dioxide Emissions and Energy*. Brussels: Commission of the European Communities, June 1992 COM (92) 226 Final 30

9 *Fourteen Proposals for a New Earth — Policy Triad for the Environment, Economy and Energy* Communication by Hiroshi Chayama (Third Secretary, Embassy of Japan, London) Tokyo: Report by the Special Committees on Energy and Environment: Industrial Structure Council, Advisory Committee for Energy, Industrial Technology Council, 25 November 1992

10 *Green Paper on The Impact of Transport on the Environment. A Community Strategy for 'sustainable mobility'* Brussels: Commission of the European Communities, Feb 1992 COM (92) 46

11 *Proposal for a Council Directive on excise duties on motor fuels from agricultural sources* Brussels: Commission of the European Communities, Feb 1992 COM (92) 36. Final

12 John Houghton *Global warming. The complete briefing* Oxford: Lion Publishing, 1994

13 *In trust for tomorrow. Report of the Labour Party Commission on the environment.* London: Labour Party, 1994

14 *Proposal for a Council Decision on the signing of the European Energy Charter Treaty and its provisional application by the European Community* Brussels: Commission of the European Communities, September 1994 COM (94) 405 Final/2

15 'Council Decision of 13 September 1993 concerning the promotion of renewable energy sources in the Community (ALTENER programme)' Brussels: Commission of the European Communities, 18 September 1993. *Official Journal of the European Communities* p.L 235

16 'Council Directive 93/76/EEC of 13 September 1993 to limit carbon dioxide emissions by improving energy efficiency (SAVE)' Brussels: Commission of the European Communities, 23 September 1993 *Official Journal of the European Communities* p.L 237

17 'Council Regulation (EEC) No 2008/90 of 29 June 1990 concerning the promotion of energy technology in Europe (THERMIE programme)' Brussels: Commission of the European Communities, 29 June 1990 *Official Journal of the European Communities* p.L185/1

18 *New and renewable energy:future prospects in the UK* London: HMSO, March 1994 Energy Paper no 62

19 I Fells, N Lucas 'Coal, the community and the environment' *Energy Focus (Journal of Energy Studies in Parliament)* 10(1) Apr 1993 pp.1-17

20 I Fells, N Lucas 'UK energy policy post privatization' *Energy Policy* 20(5) May 1992 pp.386-389. Also I Fells 'Energy strategy and the environment' *Process Safety and Environmental Protection* 70(B2) May 1992 pp.93-98

21 For example J E Lovelock *Gaia* Oxford: OUP, 1979 and *The ages of Gaia* Oxford: OUP, 1988. Quoted in ref 12 above

22 *Sustainable development. The UK strategy.* London: HMSO, Jan 1994 Cmnd. 2426

23 *Pocket guide to policy.* London: Liberal Democrat Party, 1992 ISBN 1851871292

24 *Energy efficiency in buildings* London: Environment Committee/HMSO, 1993

Organisations Connected with Energy

Amalgamated Engineering and Electrical Union
110 Peckham Road
London SE15 5EL
Telephone 0171 703 4231
Fax 0171-701 7862

Association for the Conservation of Energy
9 Sherlock Mews
London W1M 3RH
Telephone 0171-935 1495
Fax 0171-935 8346

Association of British Solid Fuel Appliance Manufacturers
77 Renfrew Street
Glasgow G2 3BZ
Telephone 0141-332 0826
Fax 0141-332 5788

Association of Electrical Contractors of Ireland
16 Main Street
Blackrock
Co. Dublin
Republic of Ireland
Telephone 353 (1) 288 6499
Fax 353 91) 288 5870

Association of Electrical Machinery Trades
Ferns Farmhouse
Great Hampdon
Great Missenden
Bucks HP16 9RG
Telphone 01494 488834
Fax 01494 1094

Association of Gas and Fluids Transmission Control Engineers
75 Benshaw Grove
Thornton Heath
Surrey CR7 8DX
Telephone 0181-771 1192
Fax 0181 771 7553

Association of Independent Electricity Producers
41 Whitehall
London SW1A 2BX
Telephone 0171- 930 9390
Fax 0171-930 9391

Association of Managerial Electrical Executives
Hayes Court
West Common Road
Hayes
Bromley BR2 7AU
Telephone 0181-462 7755

Association of Manufacturers Allied to the Electrical & Electronic Industry (AMA)
Electronic Industry (AMA)
8 Leicester Street
London WC2H 7BN
Telephone 0171-437 0678
Fax 0171-437 4901

Association of Manufacturers of Domestic Electrical Appliances
8 Leicester Street
London WC2H 7BN
Telephone 0171-437 0678
Fax 0171-437 4901

Association of Short-Circuit Testing Authorities (ASTA) Certification Services
8 Leicester Street
London WC2H 7BN
Telephone 0171-437 0678
Fax 0171-437 6089

BEAMA Capacitor Manufacturers Association
8 Leicester Street
London WC2H 7BN
Telephone 0171-437 0678
Fax 0171-437 4901

BEAMA Interactive and Mains Systems Association
8 Leicester Street
London WC2H 7BN
Telephone 0171-437 0678
Fax 0171-437 6089

BEAMA Ltd; Federation of British Electrotechnical & Allied Manufacturers Associations
8 Leicester Street
London WC2H 7BN
Telephone 0171-437 0678
Fax 0171-872 6243

British Anaerobic & Biomass Association
PO Box 7
Southend
Reading RG7 6AZ

British Combustion Equipment Manufacturers Association
The Fernery
Market Place
Midhurst
West Sussex GU29 9DP
Telephone 01730 812782
Fax 01730 813366

British Compressed Gases Association
14 Tollgate
Eastleigh
Hants SQ5 3TG
Telephone 01703 641488
Fax 01703 641477

British Energy Association
34 St James's Street
London WIH 1HD
Telephone 0171-930 1211
Fax 0171-925 0452

British Institute of Energy Economics
37 Woodville Gardens
London W5 2LL
Telephone 0181-997 3707
Fax 0181-566 7674

British Photovoltaic Society
The Warren
Bromshill Road
Eversley
Hants RG27 0PR
Telephone 01734 730820

British Radio and Electronic Equipment Manufacturers Association
19 Charing Cross Road
London WC2H 0ES
Telephone 0171-930 3206
Fax 0171-839 4613

Buildings Energy Efficiency Confederation
9 Sherlock Mews
London W1M 3RH
Telephone 0171-935 1495
Fax 0171-935 8346

Chartered Institution of Building Services Engineers
222 Balham High Road
London SW12 9BS
Telephone 0181-675 5211
Fax 0181-675 5449

Coal Authority
200 Lichfield Lane
Berry Hill
Mansfield
Notts NG18 4RG

Combined Heat and Power Association
35-37 Grosvenor Gardens
London SW1W 0BS
Telephone 0171-828 4077
Fax 0171-828 0310

Council for the Registration of Gas Installers
4 Elmwood
Chineham Business Park
Crockford Lane
Basingstoke
Hants RG24 0WG
Telephone 01256 707060
Fax 01256 708144

Decorative Gas Fire Manufacturers Association
10 Fremantle Business Centre
Millbrook Road East
Southampton S01 0JR
Telephone 01703 631593
Fax 01703 634 497

Department of Trade and Industry
123 Victoria Street
London SW1E 6RB
Telephone 0171-215 5000
Fax 0171-215 5665

Electric Trace Heating Industries Council
PO Box 28
Harpenden
Herts AL5 5UW
Telephone 01582 762843

Electrical and Engineering Staff Association
Hayes Court
West Common Road
Bromley BR2 7AU
Telephone 0181-462 7755

Electrical Contractors Association
34 Palace Court
London W2 4HY
Telephone: 0171-229 1266

Electrical Contractors Association Scotland
Bush House
Bush Estate
Midlothian EH26 08B
Telephone 0131-445 5577
Fax 0131-445 5548

Electrical Electronic Telecommunication and Plumbing Union
Hayes Court
West Common Road
Bromley BR2 7AU
Telephone 0181-462 7755
Fax 0181-462 4959

Electrical Industries Federation of Ireland
16 Main Street
Blackrock
Co. Dublin
Republic of Ireland

Electrical Installation Equipment Manufacturers Association
8 Leicester Street
London WC2H 7BN
Telephone 0171-437 0678
Fax 0171-437 4901

Electrical Power Engineers Association
Flaxman House
Gogmore Lane
Chertsey
Surrey KT16 9JS
Telephone 01932 564131
Fax 01932 567707

Electricity Arbitration Association
5 Meadow Road
Gt Gransden
Sandy
Beds SG19 3BD
Telephone 01767 677043
Fax 01767 677603

Electricity Association Services
30 Millbank
London SW1P 4RD
Telephone 0171-344 5700
Fax 0171-931 0356

Electro-Technical Council Ireland
Parnell House
Harold's Cross
Dublin 6
Republic of Ireland
Telephone 353 (1) 545819
Fax 353 (1) 545821

Energy Efficiency Office
Thames House
South Palace Street
London SW1E 5HE
Telephone 0171-238 3000

Energy Industries Council
48 Notting Hill Gate
London W11 3LQ
Telephone 0171-221 2043
Fax 0171-221 8813

Energy Saving Trust
11 Buckingham Gate
London SW1E 6LB
Telephone 0171-931 8401

Energy Systems Trade Association
PO Box 16
Stroud
Gloucestershire GL6 9YB
Telephone 01453 885226
Fax 01453 886776

Energy Technology Support Unit
Harwell
Oxon
OX11 0RA
Telephone 01235 82100

Export Group Constructional Industries
15-17 King Street
London SW1Y 6QU
Telephone 0171-930 5377
Fax 0171-930 2306

Federation of Electronic and Information Industries [Ireland]
Confederation House
Kildare Street
Dublin 2
Republic of Ireland

FEI - Federation of Electronics Industries
10-12 Russell Square
London WC1B 5EE
Telephone 0171-331 2000
Fax 0171-331 2040

Institute of Energy
18 Devonshire Street
London W1N 2AU
Telephone 0171-580 0008

Institution of Civil Engineers
1-7 Gt George Street
London SW1P 3AA
Telephone 0171-222 7722
Fax 0171-222 7500

Institution of Electrical Engineers
Savoy Place
London WC2R 0BL
Telephone 0171-240 1871
Fax 0171-240 7735

Institution of Electronics and Electrical Incorporated Engineers
Savoy Hill House, Savoy Hill
London WC2R 0BS
Telephone 0171-836 3357

Institution of Gas Engineers
21 Portland Place
London W1N 3AF
Telephone 0171-636 6603

Institution of Mechanical Engineers
1 Birdcage Walk
London SW1H 9JJ
Telephone 0171-222 7899

Institution of Mining Electrical and Mechanical Engineers
60 Silver Street
Doncaster DN1 1HT
Telephone 01302 360104
Fax 01302 730399

International Consumer Electronics Association
19 Charing Cross Road
London WC2H 0ES
Telephone 0171-930 3206
Fax 0171-437 4901

Irish Gas Association
c/o Bord Gas Eivan
PO Box 51
Inchero, Little Island
Co. Cork
Republic of Ireland
Telephone 353 (1) 509199
Fax 353 (1) 353487

Irish Transformer Manufacturers Association
Confederation House
Kildare Street
Dublin 2
Republic of Ireland
Telephone 353 (1) 677 9801
Fax 353 (1) 677 7823

National Association of Solid Fuel Wholesalers
77 Renfrew Street
Glasgow G2 3BZ
Telephone 0141-332 0826
Fax 0141-332 5788

National Energy Efficiency Association
c/o Institute of Energy
18 Devonshire Street
London W1N 2AU

National Industrial Fuel Efficiency Service
Sinderland Road
Altrincham
Cheshire WA14 5HQ
Telephone 0161-928 5791

National Joint Utilities Group
30 Millbank
London SW1P 4RD
Telephone 0171-344 5720
Fax 0171-344 5989

Offshore Manufacturers and Constructors Association
1 Melville Crescent
Edinburgh EH3 7HW
Telephone 0131 226 2470
Fax 0131-226 2471

Overhead Transmission Line Contractors Association
c/o Eve Group Plc
Minster House
Plough Lane
London SW17 QAZ
Telephone 0181-946 3085
Fax 0181-946 2156

Power Generation Contractors Association
8 Leicester Street
London WC2H 7BN
Telephone 0171-437 0678
Fax 0171-437 4901

Power Supply Manufacturers Association
8 Leicester Street
London WC2H 7BN
Telephone 0171-437 0678
Fax 0171-437 4901

Printed Circuit Interconnection Federation
399-401 Strand
London WC2R 0LT
Telephone 0171-497 1090

Resource Use Institute
Leewood
Leewood Road
Dunblane
Perthshire FK15 0DR
Telephone 01786 822161

Scottish Solar Energy Group
Department of Architecture
University of Edinburgh
Chambers Street
Edinburgh EH1 1JZ
Telephone 0131-650 2300

Society of British Gas Industries
36 Holly Walk
Royal Leamington Spa
Warwickshire CV32 4LY
Telephone 01926 334357
Fax 01926 450459

Society of Chief Electrical and Mechanical Engineers (Local Government)
County Hall
Trowbridge
Wiltshire BA15 8JA
Telephone 01225 713000
Fax 01225 713991

Solar Energy Society
19 Franklin Road
Birmingham B30 2HE
Telephone 0121-459 1248
Fax 0121-459 8206

Solar Energy Society of Ireland
Department of Electronic and Electrical Engineering
University College
Bellfield
Dublin 4
Republic of Ireland
Telephone 353(1) 269 3244
Fax 353(1) 283 0921

Solar Trade Association
Pengillon, Lerryn
Lostwithiel
Cornwall PL22 0QE
Tel 01208 873518
Fax 01208 873518

Solid Fuel Association
Victoria House
Southampton Row
London WC1B 4DH
Telephone 0171-831 5181

UK Offshore Operators Association
3 Hans Crescent
London SW1X 0LN
Telephone 0171-589 5255
Fax 0171-589 8961

UK Onshore Operators Group
63 Duke Street
London W1M 5DH
Telephone 0171-355 3393
Fax 0171-355 3704

Glossary

ANTHROPOGENIC GASES - Gases generated by human activity.

AQUIFER - Rock formation containing water in recoverable quantities.

BIOFUELS - Solid, liquid or gaseous fuels derived from crops that may be grown or used to produce energy. This includes agricultural and forestry wastes.

BIOMASS - Organic matter, from animals or plants. It may be harvested as a source of biofuels.

BIOREMEDIATION - Using microorganisms sprayed on to spilt oil in order to decompose it chemically (biodegradation).

CIRCULATING FLUIDISED BED COMBUSTORS - Solids ground down into fine particles can be made to behave rather like a liquid by lifting and agitating them in a stream of liquid or gas. Coal can be burnt in such a fluidised bed, continuously feeding in coal and tapping off the ash produced. Fluidised bed systems may be designed to operate with a defined surface or in a turbulent cloud that fills the primary combustor and circulates.

COMBINED CYCLE GENERATING PLANT - One of a variety of systems for generating electricity in more than one stage in the same plant. For example, electricity may be generated in a generator driven by a combustion turbine fired by natural gas. Heat recovered from the exhaust of the combustion turbine may be used to raise steam that drives a turbo-alternator generating further electricity.. There are many variants of this combined cycle approach.

COMBINED HEAT AND POWER - The heat rejected from a conventional generating plant may be used for industrial or district heating. Terminology is not standardised but this may be known as Combined Heat and Power or Cogeneration.

CRACKING - Chemically breaking down larger molecules such as the heavier crude oil fractions to lighter, more valuable fractions by heat, pressure and the use of catalysts.

CRITICAL MASS - The minimum mass of fissionable material that can sustain a chain reaction.

DEMAND SIDE MANAGEMENT - Schemes, originally devised in the USA, to restrict energy demand, including such measures as promoting energy-efficient lighting, improving house insulation, promoting in industry energy audits.

DISTRICT HEATING SCHEMES - Schemes where heat , usually in the form of hot water, is supplied from one source to a district or a group of buildings.

ECOSYSTEM - A system of interdependent, interacting organisms and their physical environment.

ENERGY INTENSITY - The ratio of energy used, within a country or region, to the Gross Domestic Product at constant prices.

ENRICHMENT - Processing of natural uranium to increase the proportion of fissile U235 from the naturally occurring level of about 0.7% usually to about 3% for use in thermal power reactors.

EXTERNALITIES - The environmental costs of using fossil fuels that are not borne by the purchasers of the fuels. They include damage to human health, to buildings and metalwork, and the large contribution to atmospheric CO_2 and therefore to global warming.

FLUE GAS DESULPHURISATION - The process of removing sulphur dioxide (SO_2) by means of a chemical sorbent from the flue gases, usually of a power station, before they are discharged into the atmosphere. The resulting sulphur-rich products must then be disposed of in an environmentally acceptable way.

FLUIDISED BED - Solids ground down into fine particles can be made to behave rather like a liquid by lifting and agitating them in a stream of liquid or gas. Coal can be burnt in such a fluidised bed, continuously feeding in coal and tapping off the ash produced. See also Circulating fluidised bed combustors.

FLY ASH - The fine ash from the pulverized coal burnt in power stations.

FOSSIL FUEL - A natural fuel such as coal, oil or gas formed in the geological past from the decomposition of the remains of living organisms.

FUEL CELL - A device which produces an electric current directly from a chemical reaction.

GAIA HYPOTHESIS - The proposition that life or the biosphere regulates or maintains the climate and the atmospheric composition at an optimum for itself.

GEOTHERMAL POWER - The power available by drawing on the internal heat of the earth by extracting hot water from aquifers or by drilling two holes from the surface, pumping water down one of these and using the superheated water or steam returning to the surface through the other hole.

GREENHOUSE EFFECT - Incoming radiation from the sun warms the earth's surface. The energy is transmitted outwards as thermal radiation but some is absorbed by some of the gases in the atmosphere, known as the greenhouse gases, warming the atmosphere. The effect, somewhat similar to the way that greenhouse glass retains some outgoing thermal radiation, is known as the greenhouse effect.

MAGNETOHYDRODYNAMICS - A two-stage process. Coal is burnt to produce an extremely hot gas and a chemical compound is added as a "seed". This gas is passed through a strong magnetic field and electricity is tapped from the gas stream.

MtC - Abbreviation for Megatonnes of carbon. A Megatonne is a million tonnes.

Mtoe - Abbreviation for million tonnes of oil equivalent.

NUCLEAR FISSION - Disintegration of the nucleus of a heavy atom into lighter atoms. There is a loss of mass which is converted into nuclear energy.

NUCLEAR FUSION - Fusion of nuclei of lighter atoms to form atoms of new elements, notably fusion of hydrogen isotopes to form helium . There is a loss of mass which is converted into nuclear energy.

OCEAN THERMAL ENERGY CONVERSION - Using the temperature difference between the surface of tropical seas and deep water to drive power plants.

OZONE LAYER - A stratum of the upper atmosphere in which there is an appreciable concentration of ozone — the form of oxygen with three atoms per molecule, O^3 This layer protects the earth's surface from harmful ultra-violet rays. The layer is thinning and international action has been taken to phase out production of ozone-depleting substances.

PASSIVE SOLAR DESIGN - Designing buildings so as to maximise free solar gains and reduce their energy needs for heating, cooling and lighting. It includes retrofitting features such as conservatories .

PHOTOVOLTAICS - Photovoltaic materials generate direct current electrical power when they are exposed to light. Systems using these materials have no moving parts.

PINCH TECHNOLOGY - Integrating industrial processes so that waste heat carried by hot discharged products or intermediates is recovered to heat up feed materials, and extending this to total sites.

PROVED RESERVES - The quantities of fossil fuels which geological and engineering information indicate with reasonable certainty can be recovered in the future under existing economic and operating conditions.

RENEWABLE ENERGY - Energy flows that occur naturally and repeatedly in the environment and can be harnessed for human benefit. The ultimate sources are generally the sun, gravity and the earth's rotation. In practice it is customary to include the use of wastes.

TRUE BOILING POINT CURVE - The test distillation of crude oil is carried out under standard conditions. At the higher end of the boiling point range, the products would be liable to decompose. So pressure is reduced, so that the higher boiling components can be distilled at temperatures low enough to prevent them from decomposing. Then the temperatures are corrected to what would have been the equivalent at atmospheric pressure giving the True Boiling Point curve. This curve is used to estimate the yields of products that will be obtained at particular temperatures in a commercial plant.

Index

A

AEA Technology 82,83,84
Agenda 21 4
Asia Alternative Energy Unit 99
Association of Petrochemical Producers in Europe (APPE) 49
Atomic Energy Authority, UK 82
Atomic Weapons Establishment 86
Avedoere Station, combined heat and electric power 26

B

barriers to energy efficiency 121
base chemicals from oil 49
BNFL see British Nuclear Fuels plc
BP Statistical Review of World Energy 6,24
Biological Diversity, Convention on 4
bioremediation for clearing spilt oil 47
British Coal 32,34
British Gas 64
British Nuclear Fuels plc (BNFL) 83,86
British Standards 40,46

C

Campaign for Nuclear Disarmament 78
cancers in radiation workers and their children 86
cars, fuel economy 43,45,46
catalytic systems in cars 46
CCGT see combined cycle gas turbine
CEGB see Central Electricity Generating Board
Central Electricity Generating Board (CEGB) 82
chemicals from oil 47
Chernobyl 70,77
CHP see Combined heat and power

Clean Coal Technologies 26
clean technology programme 126
Climate Change, Inter-Governmental Panel on (IPCC) 4
Climate Change, UN Framework Convention on 4,125,132
Climate Change Programme, UK 125
CO_2 fixation 33
CO_2 generated by different fuels 64
Coal Authority 34
coal, Classification 22
coal combustion 27
coal extraction 27
coal gasification and combined cycle system, integrated (IGCC) 30
coal, hybrid combined cycles 32
Coal Industry Act 34
coal into gas 30,59,64
coal: origins and composition 20,21
coal preparation 27
Coal Research Establishment 32
coal, sources and reserves 23
coal, UK 20,33,34,35
Coal Research & Development Programme, UK 16
Coal Research, IEA 22
coal, using, and protecting environment 24
cogeneration 28
Cold Chain programme 103
cold fusion, nuclear 82
combined cycle gas turbine (CCGT) 120
combined heat and power (CHP) 28,119,125,141
Commons Environment Committee 145
conservation potential
  agriculture 117
  chemicals 116
  electricity 122
  energy conversion and supply 119
  iron and steel 116
  motor drive systems 120

residential and commercial 117
transport 118
trends and areas of research 122
UK trends 124
Convention on Biological Diversity 4
Convention on Climate Change, UN Framework 4,125
cracking 41
crops as fuel 93
'crude oil' 39,40
crude oil to useful products, processing 40

D

decommissioning of nuclear reactors 76,84
Defence Radiological Service 86
Demand Side Management (DSM) 122
Department of the Environment 84
Department of Trade and Industry 84
dry wastes as a fuel 93
DSM see Demand Side Management

E

"Earth Summit", the 3,132
ecolabels 124
Economic Commission for Europe, (ECE),
coal output forecast 24
international coal classification system 23
international coal codification system 23
EEO see Energy Efficiency Office
electricity
conservation potential 122
fast-growing energy sector 9
gas-fired generation 59
electricity, UK 34
Energy Agency 144
Energy, BP Statistical Review of World 6

Energy Conservation Bill 127,L144
energy consumption per head 2
energy consumption per head, changes in 2
Energy Efficiency Office (EEO) 16,121,125,126,141
energy intensity
general 8
and conservation 114
energy, patterns of supply and demand 9
Energy Saving Trust 122,125,126,141
Energy Technology Support Unit (ETSU) 101
energy, world trends 6
Environment and Development, UN Conference on 3
Environment and Development, Rio Declaration on 4
environment and energy 3
environment, using coal and protecting 24
environmental aspects, renewables 98
environmental effects, oil 45
environmental issues, nuclear 77
Environmental Pollution, Royal Commission on 145
ETSU see Energy Technology Support Unit
EU see European Union
European Commission
and fossil fuel levy 84
carbon/energy tax, proposed 124,140
European energy policy 13
Global Photovoltaic Action Plan 103
survey of CHP schemes 125
European Fast Reactor Associates 73
European Union (EU)
ecolabels 124
energy policy 138
European Energy Charter 139
Programmes JOULE, THERMIE, ALTENER,SAVE 123,140
trends in energy pattern 12
trends in renewables 104
'externalities' and energy efficiency 121

F

fast reactors 72
Fells, Professor Ian 144
FGD see flue gas desulphurisation
fission nuclear reaction 71
flue gas desulphurisation (FGD) 25,26,28
fluidised bed combustion 28,29,30
forests, Statement of Principles on managing 4
fossil fuel levy 84
Fraunhofer Institute for Solar Energy Systems 96
Friends of the Earth 78
fuel cells 32,67,123
fusion nuclear reaction 71,80

G

Gaia hypothesis 147
gas, environmental issues 63
gas flaring 60
gas for generating electricity 59
gas for vehicles 59
gas liquids 57
gas, origins and reserves 56
gas storage 60
gas transport 61
gas, UK trends 64
gas, world trends of production, consumption, trade 58
gasification 30
gasification and combined cycle system, integrated coal (IGCC) 30
gasification, Underground 31
geothermal aquifers 95
geothermal hot dry rock 95
greenhouse effect 4
Global Photovoltaic Action Plan 103
Greenpeace 78

H

Hadley Centre for Climate Change Prediction and Research 16
HEU see highly enriched uranium
highly enriched uranium (HEU) 75
Hohenheim University 97
Houghton, John 147
hybrid combined cycles, coal 32
hydro power 94

I

IAEA see International Atomic Energy Agency
ICRP see International Commission on Radiological Protection
IEA Coal Research 22,26,33,123
IGCC see integrated coal gasification and combined cycle system
IGFC see Integrated Gasification Fuel Cell
IMO see International Maritime Organisation
India Renewable Resources Development Project 99
Institute of Petroleum, test methods 40
Integrated Coal Gasification and Combined Cycle System (IGCC) 30
Integrated Gasification Fuel Cell (IGFC) 123
Inter-Governmental Panel on Climate Change (IPCC) 4
Intermediate Technology Group and renewables 104
International Atomic Energy Agency (IAEA) 76,78
International Commission on Radiological Protection (ICRP) 85
International Maritime Organisation (IMO) 47
International Organisation for Standardization (ISO) 23
International Thermonuclear Experimental Reactor (ITER) 81
IPCC see Inter-Governmental Panel on Climate Change
ISO see International Organisation for Standardization
ITER see International Thermonuclear Experimental Reactor

J

JET see Joint European Torus
Joint European Torus (JET) 81

L

Labour Party policy 143
lead in gasoline (petrol) 45,52
leukaemia 86
Liberal Democrat policy 144
liquefied natural gas (LNG) 57,61,63
liquefied petroleum gas (LPG) 57
LNG see liquefied natural gas
Lovelock, J E 147
LPG see liquefied petroleum gas
Lucas, Professor N 144

M

magnetohydrodynamics (MHD) 32
marine pollution by oil 46
MARPOL 47
Maximum Practicable Resource, renewables 109
methane as greenhouse gas 26
MHD see magnetohydrodynamics
micro-organisms 'farmed' on oil 51
Ministry of International Trade and Industry (MITI), Japanese 137
MITI see Ministry of International Trade and Industry , Japanese
Mixed Oxide fuel (MOX) 75
MOX see Mixed Oxide fuel
municipal solid waste, as fuel 93

N

National Coal Board (NCB) 31,33
National Consumer Council 83
National Grid Co 83
National Power 83
National Radiological Protection Board (NRPB) 84,85
National Registry for Radiation Workers 86
natural gas see also gas
natural gas, composition 56

natural gas liquids (NGL) 39,57
NCB see National Coal Board
'New Earth', Japanese proposals for a 137
NFFO see Non-Fossil Fuel Obligation
NGL see natural gas liquids
NIREX see Nuclear Industry Radioactive Waste Executive
Non-Fossil Fuel Obligation (NFFO) 16,106,110
Non-Proliferation Treaty (NPT) 78
North Sea oil 52
NPT see Non-Proliferation Treaty
NRPB see National Radiological Protection Board
Nuclear Electric 83,84,86
nuclear energy
  environmental and safety issues 77
  UK Trends 82
nuclear fuels, reserves and waste disposal 74
nuclear fusion 80
Nuclear Industry Radioactive Waste Executive (NIREX) 76,83
nuclear reactions 70
nuclear reactors:
  capital requirements 73
  decommissioning 76

O

occupational exposure of radiation workers 86
Ocean Thermal Energy Conversion (OTEC) 97
OFGAS 65
oil, chemicals from 47
oil, environmental effects 45,52
oil, 'farming' micro-organisms on 51
oil, origins and composition 38
oil, UK trends 52
oil; world patterns of production, consumption and trade 42
OPEC see Organisation of Petroleum Exporting Countries

Organisation of Petroleum Exporting Countries (OPEC) 33,44
OTEC see Ocean Thermal Energy Conversion

P

Petrochemical Producers in Europe, Association of (APPE) 49
petroleum see oil
photoconversion: electrochemical photovoltaic cells 95
photoconversion: photobiological processes 95
photovoltaics 95
Pinch Technology 116
Plan for Coal 33
'Policy Triad', Japanese 137
population 6, 7
PowerGen 83
Power Save Project, Holyhead 122
price subsidies criticised 9
PSD see Solar Design, Passive

R

radiation workers, occupational exposure of 86
radiation, exposure of the public 84
REAG see Renewable Energy Advisory Group
Renewable Energy Advisory Group 92,93,105
Renewable Energy Obligation (REO) 106
renewables,
  defined 92
  Environmental Aspects 98
  in developing countries 102
  UK trends 105
  UN study 101
  World Energy Council study 101
  world trends 101
REO see Renewable Energy Obligation
Reserves, coal 11, 23
Reserves, gas 11, 57

reserves, nuclear 74
reserves, oil 11
Royal Commission on Environmental Pollution 145
Royal Society
on greenhouse effect 4
on radioactive waste 76,83,84

S

safety issues, nuclear energy 77
Sasol 27
Science Summit on population 8
Scottish Hydro-Electric 83
Scottish Nuclear Ltd 83
Scottish Power 83
Seyler, coal classification 22,23
solar heating, active 96
solar design, passive (PSD) 96
solar ponds 97
solar systems, improved 97
'Sustainable Development', UK government report 5

T

taxation, and energy efficiency 121
TBP see True Boiling Point curve, crude oil
tele-conference systems 119
thermal oxide reprocessing plant (THORP) 83
THORP see thermal oxide reprocessing plant
Three Mile Island 70
tidal current turbine 98
tidal power 94
transport 9,38,52,118,145
True Boilng Point curve (TBP), crude oil 40

U

UK Atomic Energy Authority 86
UK government reports
'Coal for the Future' 36
'Sustainable Development' 5,16
'Global Climate Change' 126,129
UK government policy
Building Regulations 127
Climate Change Programme 125
energy 16,141
renewables 106
UK Trends
coal 33
conservation 124
gas 64
oil 52
renewables 105
total energy 15
underground gasification 31
Uranium Institute survey 75
United Nations (UN)
Conference on Environment and Development 3
Environment Programme 5,99
Food and Agriculture Organisation 99
Framework Convention on Climate Change 4
International Conference on Population and Development 8
International Conference on the Peaceful Uses of Atomic Energy 78
study on renewables 101

W

wave energy 94
WEC, see World Energy Council
wet wastes as a fuel 93
WHO see World Health Organisation
wind energy 94

World Association of Nuclear Operators 77
World Bank 98
  on price subsidies 9
World Commission on Environment and Development 101
World Energy Council (WEC) 6,93
  on price subsidies 9
  recommendations 137
  study on conservation 115
  study on renewables 101
  the four cases of possible energy demand 132
World Health Organisation (WHO), Cold Chain programme 103
World Meteorological Organisation 5
world trends
  in energy 6
  nuclear 78
  renewables 101
world population 6, 7

Printed by Hobbs the Printers Ltd, Totton, Hampshire SO40 3YS